Engineering Mathematics: A Formula Handbook

N.B. Singh

DEDICATION

To Nature,

I dedicate this book to you, the source of all life. You are my inspiration, my teacher, and my friend.

Thank you for teaching me about the beauty of the world around me. Thank you for showing me the power of the natural world. Thank you for giving me a sense of peace and tranquillity.

I promise to do my part to protect you and your many wonders. I will teach my children about the importance of conservation and sustainability. I will work to make the world a better place for all living things.

Thank you for everything, Nature.

With love,

N.B Singh

Contents

1 **Linear Algebra**   1

   1.1 Vector Spaces and Subspaces   1

       1.1.1 Vector Spaces: Fundamental Properties   1

       1.1.2 Subspaces: A Closer Look   2

       1.1.3 Example: Null Space of a Matrix   2

       1.1.4 Basis and Dimension   2

       1.1.5 Linear Transformations and Image   3

       1.1.6 Eigenvalues and Eigenvectors   3

       1.1.7 Example: Eigenvalues of a Matrix   3

       1.1.8 Orthogonality and Inner Product Spaces   3

   1.2 Linear Dependence and Independence   4

       1.2.1 Definition of Linear Independence   4

       1.2.2 Definition of Linear Dependence   4

       1.2.3 Geometric Interpretation   4

       1.2.4 Testing for Linear Independence   5

       1.2.5 Numerical Example: Linear Independence Test   5

       1.2.6 Applications in Engineering   5

       1.2.7 Basis of a Vector Space   5

       1.2.8 Rank and Nullity   6

       1.2.9 Example: Rank-Nullity Theorem   6

   1.3 Matrices and Determinants   6

1.3.1   Matrix Notation and Definition 6

1.3.2   Matrix Addition and Scalar Multiplication 6

1.3.3   Matrix Multiplication 7

1.3.4   Example: Matrix Multiplication 7

1.3.5   Determinants of Matrices 7

1.3.6   Example: Determinant Calculation 8

1.3.7   Inverse of a Matrix 8

1.3.8   Example: Matrix Inversion 8

1.3.9   Eigenvalues and Eigenvectors 8

1.3.10   Example: Eigenvalues of a Matrix 8

1.3.11   Applications in Engineering 9

1.3.12   Matrix Transposition 9

1.3.13   Example: Matrix Transposition 9

1.3.14   Orthogonal Matrices 9

1.3.15   Example: Orthogonal Matrix 9

1.4   Eigenvalues and Eigenvectors 10

1.4.1   Definition of Eigenvalues and Eigenvectors 10

1.4.2   Characteristic Equation 10

1.4.3   Example: Eigenvalues of a 2×2 Matrix 10

1.4.4   Numerical Example: Eigenvalues of a 3×3 Matrix 11

1.4.5   Diagonalization of Matrices 11

1.4.6   Example: Diagonalization 11

1.4.7   Applications in Physics 11

1.4.8   Eigenvalues in Dynamic Systems 11

1.4.9   Example: Stability Analysis 12

1.4.10   Singular Value Decomposition (SVD) 12

1.4.11   Example: Singular Value Decomposition 12

1.4.12   Power Iteration for Eigenvalues 12

1.4.13   Example: Power Iteration 12

1.5   Rank of a Matrix 12

1.5.1   Definition of Rank 13

1.5.2 Row-Reduced Echelon Form 13

1.5.3 Example: Calculating Rank 13

1.5.4 Rank-Nullity Theorem 13

1.5.5 Example: Rank-Nullity Theorem 13

1.5.6 Applications in Linear Systems 14

1.5.7 Example: Solvability of a System 14

1.5.8 Matrix Inversion and Full Rank 14

1.5.9 Example: Matrix Inversion 14

1.5.10 Rank in Eigenvalue Problems 15

1.5.11 Example: Eigenvalues and Rank 15

1.5.12 Singular Value Decomposition (SVD) 15

1.5.13 Example: SVD and Rank 15

1.6 System of Linear Equations 16

1.6.1 Definition of a System of Linear Equations 16

1.6.2 Matrix Form of a System 16

1.6.3 Methods for Solving Systems 16

1.6.4 Example: Solving a System Using Gaussian Elimination . 17

1.6.5 Example: Solving a System Using Matrix Inversion 17

1.6.6 Example: Solving a System Using Cramer's Rule 18

1.6.7 Types of Solutions 18

1.6.8 Example: Unique Solution 18

1.6.9 Example: No Solution 19

1.6.10 Example: Infinite Solutions 19

1.6.11 Applications in Engineering 19

1.6.12 Example: Electrical Circuit Analysis 19

2 Calculus **21**

2.1 Functions of Single Variable 21

2.1.1 Definition of a Function 21

2.1.2 Types of Functions 21

2.1.3 Example: Polynomial Function 22

2.1.4 Example: Exponential Function 22

2.1.5 Limits of Functions 22

2.1.6 Example: Finding a Limit 22

2.1.7 Continuity of Functions 22

2.1.8 Example: Checking Continuity 23

2.1.9 Derivatives of Functions 23

2.1.10 Example: Finding Derivative 23

2.1.11 Applications of Derivatives 23

2.1.12 Example: Tangent Line 24

2.1.13 Example: Optimization 24

2.1.14 Integration of Functions 24

2.1.15 Example: Indefinite Integral 24

2.1.16 Example: Definite Integral 24

2.1.17 Fundamental Theorem of Calculus 24

2.1.18 Example: Applying Fundamental Theorem 24

2.1.19 Applications in Engineering 25

2.1.20 Example: Engineering Application 25

2.2 Limit, Continuity, and Differentiability 25

2.2.1 Limits of Functions 25

2.2.2 Example: Calculating a Limit 25

2.2.3 Continuity of Functions 25

2.2.4 Example: Checking Continuity 26

2.2.5 Differentiability of Functions 26

2.2.6 Example: Finding Derivative 26

2.2.7 Limits Involving Infinity 26

2.2.8 Example: Limit at Infinity 26

2.2.9 Continuity on Closed Intervals 26

2.2.10 Example: Continuity on Closed Interval 26

2.2.11 Differentiability Implies Continuity 27

2.2.12 Example: Differentiability vs Continuity 27

2.2.13 L'Hôpital's Rule 27

2.2.14 Example: Applying L'Hôpital's Rule 27

2.2.15 Continuity on Open Intervals 27

2.2.16 Example: Continuity without Differentiability 27

2.2.17 Intermediate Value Theorem 27

2.2.18 Example: Applying Intermediate Value Theorem 28

2.2.19 Mean Value Theorem 28

2.2.20 Example: Applying Mean Value Theorem 28

2.2.21 Applications in Engineering 28

2.2.22 Example: Engineering Application 28

2.3 Mean Value Theorems 28

2.3.1 Rolle's Theorem 29

2.3.2 Example: Applying Rolle's Theorem 29

2.3.3 Mean Value Theorem 29

2.3.4 Example: Applying Mean Value Theorem 29

2.3.5 Cauchy's Mean Value Theorem 29

2.3.6 Example: Applying Cauchy's Mean Value Theorem 30

2.3.7 Applications in Calculus 30

2.3.8 Example: Analyzing Functions 30

2.3.9 Generalization to Higher Derivatives 30

2.3.10 Example: Higher Derivatives 30

2.3.11 Cautionary Note on Assumptions 30

2.3.12 Example: Checking Assumptions 30

2.3.13 Extensions to Vector-Valued Functions 31

2.3.14 Example: Vector-Valued Function 31

2.4 Indeterminate Forms and L'Hôpital's Rule 31

2.4.1 Indeterminate Forms 31

2.4.2 Example: Evaluating $\frac{0}{0}$ 31

2.4.3 L'Hôpital's Rule 32

2.4.4 Example: Applying L'Hôpital's Rule 32

2.4.5 Extended L'Hôpital's Rule 32

2.4.6 Example: Applying Extended L'Hôpital's Rule 32

2.4.7   Handling $\infty - \infty$ Forms   32

2.4.8   Example: Handling $\infty - \infty$ 32

2.4.9   Indeterminate Forms and Trigonometric Limits   33

2.4.10 Example: Trigonometric Limit   33

2.4.11 Application to Improper Integrals   33

2.4.12 Example: Improper Integral   33

2.4.13 Cautionary Note on Applying L'Hôpital's Rule   33

2.4.14 Example: Misapplication of L'Hôpital's Rule   33

2.4.15 Applications in Engineering   34

2.4.16 Example: Engineering Application   34

2.5   Taylor's Theorem   34

2.5.1   Taylor's Theorem Statement   34

2.5.2   Taylor Series   34

2.5.3   Maclaurin Series   35

2.5.4   Example: Maclaurin Series for $\sin(x)$ 35

2.5.5   Error in Taylor Approximation   35

2.5.6   Example: Estimating Error   35

2.5.7   Convergence of Taylor Series   35

2.5.8   Example: Testing Convergence   36

2.5.9   Interval of Convergence   36

2.5.10 Example: Determining Interval of Convergence   36

2.5.11 Multivariable Taylor Series   36

2.5.12 Example: Multivariable Taylor Series   36

2.5.13 Applications in Engineering   36

2.5.14 Example: Engineering Application   36

2.5.15 Taylor Series for Common Functions   37

2.5.16 Example: Using Taylor Series   37

2.6   Maxima and Minima   37

2.6.1   Local Extrema   37

2.6.2   Critical Points   37

2.6.3   Example: Finding Critical Points   37

2.6.4 First Derivative Test 38

2.6.5 Example: Applying First Derivative Test 38

2.6.6 Second Derivative Test 38

2.6.7 Example: Applying Second Derivative Test 38

2.6.8 Absolute Extrema 38

2.6.9 Example: Finding Absolute Extrema 38

2.6.10 Global Optimization 39

2.6.11 Example: Global Optimization 39

2.6.12 Applications in Engineering Design 39

2.6.13 Example: Engineering Optimization 39

2.6.14 Lagrange Multipliers 39

2.6.15 Example: Lagrange Multipliers 39

2.6.16 Sensitivity Analysis 40

2.6.17 Example: Sensitivity Analysis 40

2.7 Integration and its Applications 40

2.7.1 Definite and Indefinite Integrals 40

2.7.2 Example: Evaluating a Definite Integral 40

2.7.3 Integration Techniques 40

2.7.4 Example: Substitution Method 41

2.7.5 Example: Integration by Parts 41

2.7.6 Applications in Physics: Work 41

2.7.7 Example: Calculating Work 41

2.7.8 Applications in Engineering: Fluid Pressure 41

2.7.9 Example: Fluid Pressure Calculation 41

2.7.10 Applications in Economics: Revenue 41

2.7.11 Example: Revenue Calculation 42

2.7.12 Improper Integrals 42

2.7.13 Example: Evaluating an Improper Integral 42

2.7.14 Applications in Probability: Cumulative Distribution

Function 42

2.7.15 Example: Probability Calculation 42

2.7.16 Applications in Signal Processing: Fourier Transform . . . 42

2.7.17 Example: Fourier Transform 43

2.7.18 Applications in Geometry: Surface Area 43

2.7.19 Example: Surface Area Calculation 43

2.8 Differential Equations 43

   2.8.1 First-Order Ordinary Differential Equations 43

   2.8.2 Example: Separation of Variables 43

   2.8.3 Example: Integrating Factor 44

   2.8.4 Second-Order Ordinary Differential Equations 44

   2.8.5 Example: Characteristic Equation 44

   2.8.6 Example: Undetermined Coefficients 44

   2.8.7 Systems of Differential Equations 44

   2.8.8 Example: Linear System 44

   2.8.9 Partial Differential Equations 44

   2.8.10 Example: Heat Equation 45

   2.8.11 Example: Wave Equation 45

   2.8.12 Numerical Methods for Differential Equations 45

   2.8.13 Example: Euler's Method 45

   2.8.14 Example: Runge-Kutta Method 45

   2.8.15 Boundary Value Problems 45

   2.8.16 Example: Sturm-Liouville Problem 46

   2.8.17 Applications in Engineering: Control Systems 46

   2.8.18 Example: Control System Dynamics 46

   2.8.19 Applications in Physics: Newton's Law of Cooling 46

   2.8.20 Example: Newton's Law of Cooling 46

3 Differential Equations   47

3.1 First-order linear and nonlinear differential equations 47

   3.1.1 First-order Linear Differential Equations 47

   3.1.2 First-order Nonlinear Differential Equations 48

   3.1.3 Applications in Engineering: RC Circuits 48

3.1.4 Applications in Biology: Population Growth 49

3.1.5 Applications in Chemistry: Reaction Kinetics 49

3.1.6 Applications in Economics: Exponential Growth 49

3.1.7 Systems of First-order Differential Equations 50

4 Differential Equations   **51**

4.1 Higher-order linear differential equations with constant coefficients 51

4.1.1 General Form of the Equation 51

4.1.2 Applications in Mechanical Engineering: Vibrations . . . 53

4.1.3 Applications in Electrical Engineering: LRC Circuits . . . 53

4.1.4 Applications in Physics: Newton's Law of Motion 53

4.1.5 Numerical Solutions 53

4.2 Cauchy's and Euler's Equations 54

4.2.1 Cauchy's Equation 54

4.2.2 Euler's Equation 55

4.2.3 Cauchy-Euler Equation 55

4.3 Laplace Transforms 56

4.3.1 Definition of Laplace Transform 56

4.3.2 Properties of Laplace Transform 57

4.3.3 Inverse Laplace Transform 57

4.3.4 Solving Differential Equations 58

4.3.5 Applications in Control Systems: Transfer Functions . . . 58

4.3.6 Applications in Electrical Circuits: Response Analysis . . 59

4.4 Partial Differential Equations and Their Solutions 59

4.4.1 Classification of PDEs 59

4.4.2 Method of Separation of Variables 60

4.4.3 Characteristic Curves for Nonlinear PDEs 60

4.4.4 Applications in Fluid Dynamics: Navier-Stokes Equation 61

4.4.5 Applications in Structural Mechanics: Wave Equation . . 61

5   Complex Analysis   63

    5.1   Analytic Functions   63

        5.1.1   Definition of Analytic Functions   63

        5.1.2   Properties of Analytic Functions   64

        5.1.3   Complex Integration and Cauchy's Theorem   64

        5.1.4   Singularities and Residues   65

        5.1.5   Applications in Signal Processing: Laplace Transform . .   65

    5.2   Cauchy-Riemann Equations   66

        5.2.1   Derivation of Cauchy-Riemann Equations   66

        5.2.2   Interpretation of Cauchy-Riemann Equations   66

        5.2.3   Polar Form and Cauchy-Riemann Equations   67

        5.2.4   Applications in Fluid Dynamics: Complex Potential . . .   67

        5.2.5   Numerical Examples: Solving Equations   68

    5.3   Contour Integration   68

        5.3.1   Introduction to Contour Integration   68

        5.3.2   Cauchy's Integral Formula   69

        5.3.3   Residue Theory   70

        5.3.4   Applications in Signal Processing: Fourier Transform . . .   70

        5.3.5   Numerical Examples: Complex Integrals   71

    5.4   Residue Theorem and Its Applications   71

        5.4.1   Introduction to the Residue Theorem   71

        5.4.2   Applications in Control Systems: Laplace Transform . . .   72

        5.4.3   Pole-Zero Analysis and Filter Design   72

        5.4.4   Applications in Electromagnetics: Complex Integration .   72

        5.4.5   Numerical Examples: Engineering Simulations   73

    5.5   Conformal Mappings   73

        5.5.1   Introduction to Conformal Mappings   73

        5.5.2   Properties of Conformal Mappings   74

        5.5.3   Applications in Heat Conduction: Complex Potential . . .   75

        5.5.4   Mapping of Special Regions: Upper Half-Plane   75

        5.5.5   Numerical Examples: Engineering Simulations   75

6 Probability and Statistics   **77**

6.1   Probability Space and Events 77

     6.1.1   Introduction to Probability Space 77

     6.1.2   Events and Their Properties 78

     6.1.3   Probability Calculations and Examples 79

     6.1.4   Conditional Probability 79

     6.1.5   Bayes' Theorem 80

     6.1.6   Numerical Examples: Engineering Applications 80

6.2   Random Variables and Probability Distributions 81

     6.2.1   Introduction to Random Variables 81

     6.2.2   Common Probability Distributions 81

     6.2.3   Expectation and Variance 82

     6.2.4   Sampling Distributions 83

     6.2.5   Applications in Reliability Engineering 83

     6.2.6   Numerical Examples: Engineering Applications 83

6.3   Mean, Median, Mode, and Standard Deviation 84

     6.3.1   Mean (Average) 84

     6.3.2   Median 85

     6.3.3   Mode 85

     6.3.4   Standard Deviation 85

     6.3.5   Applications in Engineering 85

     6.3.6   Working Examples 86

     6.3.7   Numerical Examples: Engineering Applications 86

6.4   Probability Density Functions 87

     6.4.1   Introduction to Probability Density Functions 87

     6.4.2   Definition of Probability Density Function 87

     6.4.3   Key Properties of PDFs 87

     6.4.4   Common Probability Density Functions 88

     6.4.5   Applications in Engineering 88

     6.4.6   Working Examples 89

6.5   Correlation and Regression Analysis 89

6.5.1 Correlation Analysis 89

6.5.2 Regression Analysis 90

6.5.3 Applications in Engineering 90

6.5.4 Working Examples 91

6.5.5 Numerical Examples: Engineering Applications 91

7 Numerical Methods   93

7.1 Solutions of Algebraic and Transcendental Equations 93

7.1.1 Bisection Method 93

7.1.2 Newton-Raphson Method 94

7.1.3 Secant Method 94

7.1.4 Applications in Engineering 94

7.1.5 Working Examples 94

7.1.6 Numerical Examples: Engineering Applications 95

7.2 Interpolation and Approximation 95

7.2.1 Linear Interpolation 95

7.2.2 Lagrange Interpolation 96

7.2.3 Least Squares Approximation 96

7.2.4 Applications in Engineering 96

7.2.5 Working Examples 97

7.2.6 Numerical Examples: Engineering Applications 97

7.3 Numerical Integration and Differentiation 97

7.3.1 Numerical Integration 98

7.3.2 Numerical Differentiation 98

7.3.3 Applications in Engineering 99

7.3.4 Working Examples 99

7.3.5 Numerical Examples: Engineering Applications 100

7.4 Solution of Ordinary Differential Equations 100

7.4.1 Euler's Method 100

7.4.2 Runge-Kutta Methods 100

7.4.3 Applications in Engineering 101

7.4.4   Working Examples 101

7.4.5   Numerical Examples: Engineering Applications 102

7.5   Finite Difference Methods 102

7.5.1   Forward Difference 102

7.5.2   Central Difference 102

7.5.3   Backward Difference 103

7.5.4   Finite Difference Scheme for Heat Equation 103

7.5.5   Applications in Engineering 103

7.5.6   Working Examples 103

7.5.7   Numerical Examples: Engineering Applications 104

8   Transform Theory   105

8.1   Laplace Transforms 105

8.1.1   Definition of the Laplace Transform 105

8.1.2   Properties of Laplace Transforms 105

8.1.3   Applications in Engineering 106

8.1.4   Working Examples 106

8.1.5   Numerical Examples: Engineering Applications 107

8.2   Fourier Series and Fourier Transforms 107

8.2.1   Fourier Series 107

8.2.2   Fourier Transform 108

8.2.3   Applications in Engineering 108

8.2.4   Working Examples 109

8.2.5   Numerical Examples: Engineering Applications 109

8.3   Z-Transforms 109

8.3.1   Definition of Z-Transform 110

8.3.2   Properties of Z-Transforms 110

8.3.3   Applications in Engineering 110

8.3.4   Working Examples 111

8.3.5   Numerical Examples: Engineering Applications 111

8.4   Convolution and Correlation 111

8.4.1   Convolution 112

8.4.2   Correlation 112

8.4.3   Properties of Convolution and Correlation 112

8.4.4   Working Examples 113

8.4.5   Numerical Examples: Engineering Applications 113

9   Linear Programming   115

9.1   Formulation of Linear Programming Problems 115

9.1.1   Introduction to Linear Programming 115

9.1.2   Formulation Steps 116

9.1.3   Working Examples 117

9.1.4   Numerical Examples: Engineering Applications 117

9.2   Basic Concepts of Graphical and Simplex Methods 117

9.2.1   Graphical Method 118

9.2.2   Simplex Method 118

9.2.3   Numerical Examples: Engineering Applications 119

9.3   Duality and Dual Simplex Method 119

9.3.1   Duality in Linear Programming 120

9.3.2   Dual Simplex Method 120

9.3.3   Numerical Examples: Engineering Applications 121

10   Probability and Statistics (Again)   123

10.1   Descriptive Statistics 123

10.1.1   Measures of Central Tendency 123

10.1.2   Measures of Dispersion 124

10.1.3   Working Examples 124

10.1.4   Numerical Examples: Engineering Applications 125

10.2   Probability Distributions 125

10.2.1   Discrete Probability Distributions 125

10.2.2   Continuous Probability Distributions 126

10.2.3   Working Examples 126

10.2.4 Numerical Examples: Engineering Applications 127

10.3 Statistical Inference 127

    10.3.1 Point Estimation 127

    10.3.2 Interval Estimation 127

    10.3.3 Hypothesis Testing 128

    10.3.4 Working Examples 128

    10.3.5 Numerical Examples: Engineering Applications 128

10.4 Hypothesis Testing 129

    10.4.1 Basics of Hypothesis Testing 129

    10.4.2 Steps in Hypothesis Testing 129

    10.4.3 Types of Errors 130

    10.4.4 Working Examples 130

    10.4.5 Numerical Examples: Engineering Applications 130

11 Graph Theory   **131**

11.1 Basic Concepts 131

    11.1.1 Graphs and Terminology 131

    11.1.2 Types of Graphs 132

    11.1.3 Working Examples 132

    11.1.4 Numerical Examples: Engineering Applications 132

11.2 Trees and Their Applications 133

    11.2.1 Definition and Properties of Trees 133

    11.2.2 Types of Trees 133

    11.2.3 Working Examples 133

    11.2.4 Numerical Examples: Optimization 134

11.3 Connectivity and Network Flows 134

    11.3.1 Connectivity in Graphs 134

    11.3.2 Network Flows 135

    11.3.3 Working Examples 135

    11.3.4 Numerical Examples 135

Preface

Welcome to *Engineering Mathematics: A Formula Handbook*. This handbook is designed to be a comprehensive reference for engineering students, professionals, and anyone seeking a quick and reliable guide to essential mathematical concepts used in various engineering disciplines.

Purpose of the Handbook

Engineering Mathematics is a vast field, and its applications are fundamental to the success of engineering projects. This handbook aims to provide a condensed and accessible compilation of key mathematical formulas, theorems, and concepts relevant to engineering.

Structure of the Handbook

The handbook is organized into chapters, each focusing on a specific branch of mathematics commonly used in engineering. From algebra and calculus to differential equations and linear programming, this handbook covers a wide range of topics. Each chapter includes concise explanations, important formulas, and practical examples to aid understanding.

Features of the Handbook

- **Comprehensive Coverage:** All major branches of mathematics relevant to engineering are covered.

- **Practical Examples:** Real-world examples illustrate the application of mathematical concepts.

- **Concise Formulas:** Formulas are presented in a clear and concise manner for quick reference.

Who Should Use This Handbook

- **Students:** Ideal for engineering students studying mathematics courses.

- **Professionals:** A handy reference for practicing engineers seeking quick solutions.

- **Educators:** Useful as a supplementary resource for teaching engineering mathematics.

Happy Learning!

I hope this handbook serves as a valuable companion in your journey through engineering mathematics. May it simplify complex concepts and enhance your understanding.

Chapter 1

Linear Algebra

1.1 Vector Spaces and Subspaces

Linear algebra provides a powerful framework for understanding vector spaces and their subspaces. A *vector space* is a set of vectors equipped with two operations: vector addition and scalar multiplication. Let's delve into the essential concepts and properties.

1.1.1 Vector Spaces: Fundamental Properties

A vector space V over a field $\mathbb{F}$ satisfies the following properties for all vectors $\mathbf{u}, \mathbf{v}, \mathbf{w} \in V$ and scalars $c, d \in \mathbb{F}$:

1. **Closure under Addition:** $\mathbf{u} + \mathbf{v} \in V$

2. **Associativity of Addition:** $\mathbf{u} + (\mathbf{v} + \mathbf{w}) = (\mathbf{u} + \mathbf{v}) + \mathbf{w}$

3. **Existence of Zero Vector:** There exists $\mathbf{0} \in V$ such that $\mathbf{u} + \mathbf{0} = \mathbf{u}$

4. **Existence of Additive Inverses:** For every $\mathbf{u} \in V$, there exists $\mathbf{v} \in V$ such that $\mathbf{u} + \mathbf{v} = \mathbf{0}$

5. **Closure under Scalar Multiplication:** $c \cdot \mathbf{u} \in V$

6. **Compatibility of Scalar Multiplication:** $c \cdot (d \cdot \mathbf{u}) = (cd) \cdot \mathbf{u}$

7. **Identity Element for Scalar Multiplication:** $1 \cdot \mathbf{u} = \mathbf{u}$

8. **Distributivity of Scalar Multiplication:** $c \cdot (\mathbf{u} + \mathbf{v}) = c \cdot \mathbf{u} + c \cdot \mathbf{v}$

9. **Distributivity of Scalar Multiplication over Field Addition:** $(c + d) \cdot \mathbf{u} = c \cdot \mathbf{u} + d \cdot \mathbf{u}$

Example: Euclidean Space $\mathbb{R}^n$

Consider $\mathbb{R}^n$ as a vector space, where vectors are n-tuples of real numbers. The vector addition and scalar multiplication operations are defined component-wise.

Numerical Example: Vector Addition in $\mathbb{R}^3$

$$\text{Let } \mathbf{u} = \begin{bmatrix} 1 \\ 2 \\ 3 \end{bmatrix} \text{ and } \mathbf{v} = \begin{bmatrix} -2 \\ 1 \\ 0 \end{bmatrix}. \text{ The sum } \mathbf{u} + \mathbf{v} \text{ is calculated as } \begin{bmatrix} -1 \\ 3 \\ 3 \end{bmatrix}.$$

1.1.2   Subspaces: A Closer Look

A *subspace* of a vector space V is a subset U of V that is itself a vector space. Subspaces inherit the vector addition and scalar multiplication operations of the parent space.

1.1.3   Example: Null Space of a Matrix

Consider a matrix A and its null space, denoted as $\text{Null}(A)$. It is the set of all solutions to the homogeneous equation $A\mathbf{x} = \mathbf{0}$.

1.1.4   Basis and Dimension

A *basis* for a vector space V is a set of linearly independent vectors that span V. The *dimension* of V is the number of vectors in any basis of V.

Example: Basis of $\mathbb{R}^3$

For $\mathbb{R}^3$, the standard basis is $\{\mathbf{i}, \mathbf{j}, \mathbf{k}\}$, where $\mathbf{i} = \begin{bmatrix} 1 \\ 0 \\ 0 \end{bmatrix}$, $\mathbf{j} = \begin{bmatrix} 0 \\ 1 \\ 0 \end{bmatrix}$, and $\mathbf{k} = \begin{bmatrix} 0 \\ 0 \\ 1 \end{bmatrix}$.

1.1.5 Linear Transformations and Image

A *linear transformation* $T : V \rightarrow W$ between vector spaces V and W preserves vector addition and scalar multiplication. The *image* of T is the set of all vectors $T(\mathbf{v})$ for $\mathbf{v} \in V$.

Example: Linear Transformation in $\mathbb{R}^2$

Consider a linear transformation $T : \mathbb{R}^2 \rightarrow \mathbb{R}^2$ defined by $T(\mathbf{v}) = A\mathbf{v}$, where A is a 2×2 matrix.

1.1.6 Eigenvalues and Eigenvectors

For a linear transformation $T : V \rightarrow V$, an *eigenvalue* λ and its corresponding *eigenvector* $\mathbf{v}$ satisfy $T(\mathbf{v}) = \lambda \mathbf{v}$.

1.1.7 Example: Eigenvalues of a Matrix

Consider a matrix A and its eigenvalues obtained by solving the characteristic equation $\det(A - \lambda I) = 0$.

1.1.8 Orthogonality and Inner Product Spaces

In an inner product space, vectors can be orthogonal, and the concept of angle between vectors is defined.

Example: Orthogonal Vectors in $\mathbb{R}^3$

Consider two vectors $\mathbf{u} = \begin{bmatrix} 1 \\ 0 \\ -1 \end{bmatrix}$ and $\mathbf{v} = \begin{bmatrix} 1 \\ 2 \\ 1 \end{bmatrix}$. These vectors are orthogonal.

1.2   Linear Dependence and Independence

Understanding the concepts of linear dependence and independence is crucial in linear algebra, providing insights into the relationships between vectors within a vector space.

1.2.1   Definition of Linear Independence

A set of vectors $\{\mathbf{v}_1, \mathbf{v}_2, \ldots, \mathbf{v}_n\}$ in a vector space V is said to be *linearly independent* if the only solution to the equation $c_1\mathbf{v}_1 + c_2\mathbf{v}_2 + \ldots + c_n\mathbf{v}_n = \mathbf{0}$ is $c_1 = c_2 = \ldots = c_n = 0$. In other words, no vector in the set can be written as a linear combination of the others.

1.2.2   Definition of Linear Dependence

Conversely, a set of vectors is *linearly dependent* if there exist constants $c_1, c_2, \ldots, c_n$, not all zero, such that $c_1\mathbf{v}_1 + c_2\mathbf{v}_2 + \ldots + c_n\mathbf{v}_n = \mathbf{0}$.

1.2.3   Geometric Interpretation

Linearly independent vectors point in different directions in space, and no vector in the set lies on the span of the others. Linearly dependent vectors, on the other hand, can be thought of as lying on a common plane or line.

Example: Linear Independence in $\mathbb{R}^2$

Consider two vectors $\mathbf{v}_1 = \begin{bmatrix} 1 \\ 0 \end{bmatrix}$ and $\mathbf{v}_2 = \begin{bmatrix} 0 \\ 1 \end{bmatrix}$ in $\mathbb{R}^2$. These vectors are linearly independent because the only solution to $c_1\mathbf{v}_1 + c_2\mathbf{v}_2 = \mathbf{0}$ is $c_1 = c_2 = 0$.

Example: Linear Dependence in $\mathbb{R}^3$

Consider three vectors $\mathbf{u} = \begin{bmatrix} 1 \\ 2 \\ 3 \end{bmatrix}$, $\mathbf{v} = \begin{bmatrix} -2 \\ 1 \\ 0 \end{bmatrix}$, and $\mathbf{w} = \begin{bmatrix} 0 \\ 3 \\ -1 \end{bmatrix}$ in $\mathbb{R}^3$. These vectors are linearly dependent since $2\mathbf{u} - \mathbf{v} + \frac{1}{2}\mathbf{w} = \mathbf{0}$.

1.2.4 Testing for Linear Independence

To determine if a set of vectors is linearly independent, one can form a matrix using these vectors and perform row operations to check if the matrix is row-equivalent to the identity matrix.

1.2.5 Numerical Example: Linear Independence Test

Consider vectors $\mathbf{a} = \begin{bmatrix} 1 \\ 2 \\ 1 \end{bmatrix}$, $\mathbf{b} = \begin{bmatrix} 2 \\ -1 \\ 3 \end{bmatrix}$, and $\mathbf{c} = \begin{bmatrix} 0 \\ 1 \\ -2 \end{bmatrix}$. Form the matrix $M = [\mathbf{a}\,\mathbf{b}\,\mathbf{c}]$, and perform row operations to determine if the vectors are linearly independent.

$$M \sim \begin{bmatrix} 1 & 2 & 0 \\ 2 & -1 & 1 \\ 1 & 3 & -2 \end{bmatrix} \sim \begin{bmatrix} 1 & 2 & 0 \\ 0 & -5 & 1 \\ 0 & 1 & -2 \end{bmatrix} \sim \begin{bmatrix} 1 & 2 & 0 \\ 0 & 1 & -\frac{1}{5} \\ 0 & 0 & 0 \end{bmatrix}$$

Since the row-echelon form has a row of zeros, the vectors are linearly dependent.

1.2.6 Applications in Engineering

Understanding linear dependence and independence is crucial in solving systems of linear equations, analyzing structures, and studying physical phenomena where vectors play a significant role.

1.2.7 Basis of a Vector Space

A set of linearly independent vectors that span a vector space V is called a *basis* for V. The number of vectors in a basis is the dimension of the vector space.

Example: Basis of $\mathbb{R}^3$

Consider vectors $\mathbf{i} = \begin{bmatrix} 1 \\ 0 \\ 0 \end{bmatrix}$, $\mathbf{j} = \begin{bmatrix} 0 \\ 1 \\ 0 \end{bmatrix}$, and $\mathbf{k} = \begin{bmatrix} 0 \\ 0 \\ 1 \end{bmatrix}$ in $\mathbb{R}^3$. This set forms the standard basis.

1.2.8   Rank and Nullity

The *rank* of a matrix is the maximum number of linearly independent rows or columns. The *nullity* is the dimension of the null space.

1.2.9   Example: Rank-Nullity Theorem

For a matrix A, the rank-nullity theorem states that $\text{rank}(A) + \text{nullity}(A) =$ number of columns of A.

1.3   Matrices and Determinants

Matrices are fundamental in linear algebra, providing a concise way to represent and manipulate linear transformations and systems of linear equations. Let's delve into the essential concepts and properties of matrices and determinants.

1.3.1   Matrix Notation and Definition

An $m \times n$ matrix A is a rectangular array of numbers with m rows and n columns. Each element a_{ij} of the matrix A is located at the intersection of the i-th row and j-th column.

$$A = \begin{bmatrix} a_{11} & a_{12} & \dots & a_{1n} \\ a_{21} & a_{22} & \dots & a_{2n} \\ \vdots & \vdots & \ddots & \vdots \\ a_{m1} & a_{m2} & \dots & a_{mn} \end{bmatrix}$$

1.3.2   Matrix Addition and Scalar Multiplication

Given matrices $A = [a_{ij}]$ and $B = [b_{ij}]$ of the same order, their sum $A + B$ and scalar product cA are defined as follows:

$$A + B = \begin{bmatrix} a_{11} + b_{11} & a_{12} + b_{12} & \dots & a_{1n} + b_{1n} \\ a_{21} + b_{21} & a_{22} + b_{22} & \dots & a_{2n} + b_{2n} \\ \vdots & \vdots & \ddots & \vdots \\ a_{m1} + b_{m1} & a_{m2} + b_{m2} & \dots & a_{mn} + b_{mn} \end{bmatrix}$$

$$cA = \begin{bmatrix} ca_{11} & ca_{12} & \dots & ca_{1n} \\ ca_{21} & ca_{22} & \dots & ca_{2n} \\ \vdots & \vdots & \ddots & \vdots \\ ca_{m1} & ca_{m2} & \dots & ca_{mn} \end{bmatrix}$$

1.3.3 Matrix Multiplication

The product of two matrices A and B is defined only if the number of columns in A is equal to the number of rows in B. The product $C = AB$ is given by:

$$C_{ij} = a_{i1}b_{1j} + a_{i2}b_{2j} + \dots + a_{in}b_{nj}$$

1.3.4 Example: Matrix Multiplication

Let $A = \begin{bmatrix} 2 & -1 \\ 3 & 4 \end{bmatrix}$ and $B = \begin{bmatrix} 5 & 2 \\ -1 & 6 \end{bmatrix}$. The product $C = AB$ is calculated as:

$$C = \begin{bmatrix} (2 \cdot 5 + (-1) \cdot (-1)) & (2 \cdot 2 + (-1) \cdot 6) \\ (3 \cdot 5 + 4 \cdot (-1)) & (3 \cdot 2 + 4 \cdot 6) \end{bmatrix} = \begin{bmatrix} 11 & -10 \\ 17 & 26 \end{bmatrix}$$

1.3.5 Determinants of Matrices

The determinant of a square matrix A, denoted as $\det(A)$, is a scalar value that can be computed from its elements. For a 2×2 matrix:

$$A = \begin{bmatrix} a & b \\ c & d \end{bmatrix}, \quad \det(A) = ad - bc$$

For a 3×3 matrix:

$$A = \begin{bmatrix} a & b & c \\ d & e & f \\ g & h & i \end{bmatrix}, \quad \det(A) = a(ei - fh) - b(di - fg) + c(dh - eg)$$

1.3.6   Example: Determinant Calculation

Consider the matrix $B = \begin{bmatrix} 3 & 1 \\ -2 & 4 \end{bmatrix}$. The determinant $\det(B)$ is calculated as $3 \cdot 4 - 1 \cdot (-2) = 14$.

1.3.7   Inverse of a Matrix

For a square matrix A, if there exists a matrix A^{-1} such that $AA^{-1} = A^{-1}A = I$, where I is the identity matrix, then A is invertible and A^{-1} is its inverse.

1.3.8   Example: Matrix Inversion

Let $C = \begin{bmatrix} 1 & 2 \\ 3 & 4 \end{bmatrix}$. The inverse of C (C^{-1}) is given by:

$$C^{-1} = \frac{1}{(1 \cdot 4 - 2 \cdot 3)} \begin{bmatrix} 4 & -2 \\ -3 & 1 \end{bmatrix} = \begin{bmatrix} -2 & 1 \\ 1.5 & -0.5 \end{bmatrix}$$

1.3.9   Eigenvalues and Eigenvectors

For a square matrix A, an *eigenvalue* λ and its corresponding *eigenvector* $\mathbf{v}$ satisfy $A\mathbf{v} = \lambda\mathbf{v}$.

1.3.10   Example: Eigenvalues of a Matrix

Consider a matrix $D = \begin{bmatrix} 2 & 1 \\ 1 & 3 \end{bmatrix}$. The eigenvalues are obtained by solving the characteristic equation $\det(D - \lambda I) = 0$.

$$\det\left(\begin{bmatrix} 2-\lambda & 1 \\ 1 & 3-\lambda \end{bmatrix}\right) = (2-\lambda)(3-\lambda) - 1 = \lambda^2 - 5\lambda + 5 = 0$$

The solutions to this quadratic equation give the eigenvalues of matrix D.

1.3.11 Applications in Engineering

Matrices and determinants play a crucial role in various engineering applications, including solving systems of linear equations, analyzing structures, and modeling dynamic systems.

1.3.12 Matrix Transposition

The transpose of a matrix A, denoted as A^T, is obtained by interchanging its rows and columns.

1.3.13 Example: Matrix Transposition

Consider $E = \begin{bmatrix} 1 & 2 & 3 \\ 4 & 5 & 6 \end{bmatrix}$. The transpose E^T is:

$$E^T = \begin{bmatrix} 1 & 4 \\ 2 & 5 \\ 3 & 6 \end{bmatrix}$$

1.3.14 Orthogonal Matrices

A square matrix Q is *orthogonal* if $Q^T Q = QQ^T = I$. Orthogonal matrices preserve lengths and angles, making them valuable in applications involving rotations and reflections.

1.3.15 Example: Orthogonal Matrix

Consider $F = \begin{bmatrix} 0 & -1 \\ 1 & 0 \end{bmatrix}$. This matrix represents a 90° counterclockwise rotation. It is orthogonal as $F^T F = FF^T = I$.

1.4   Eigenvalues and Eigenvectors

Eigenvalues and eigenvectors are fundamental concepts in linear algebra, providing valuable insights into linear transformations and matrix properties.

1.4.1   Definition of Eigenvalues and Eigenvectors

For a square matrix A, a scalar λ is an eigenvalue if there exists a non-zero vector $\mathbf{v}$ such that $A\mathbf{v} = \lambda\mathbf{v}$. The vector $\mathbf{v}$ is the corresponding eigenvector.

1.4.2   Characteristic Equation

Eigenvalues are obtained by solving the characteristic equation $\det(A - \lambda I) = 0$, where I is the identity matrix. For a 2×2 matrix A:

$$\det(A - \lambda I) = \begin{vmatrix} a - \lambda & b \\ c & d - \lambda \end{vmatrix} = (a - \lambda)(d - \lambda) - bc = 0$$

For a 3×3 matrix A:

$$\det(A - \lambda I) = \begin{vmatrix} a - \lambda & b & c \\ d & e - \lambda & f \\ g & h & i - \lambda \end{vmatrix} = 0$$

And so on for larger matrices.

1.4.3   Example: Eigenvalues of a 2×2 Matrix

Consider the matrix $B = \begin{bmatrix} 3 & 1 \\ -2 & 4 \end{bmatrix}$. The characteristic equation is:

$$\det(B - \lambda I) = \begin{vmatrix} 3 - \lambda & 1 \\ -2 & 4 - \lambda \end{vmatrix} = (3 - \lambda)(4 - \lambda) - 1 = \lambda^2 - 7\lambda + 11 = 0$$

Solving this quadratic equation gives the eigenvalues of matrix B.

1.4.4 Numerical Example: Eigenvalues of a 3×3 Matrix

Let $C = \begin{bmatrix} 1 & 0 & 0 \\ 0 & 2 & 0 \\ 0 & 0 & 3 \end{bmatrix}$. The characteristic equation is:

$$\det(C - \lambda I) = \begin{vmatrix} 1 - \lambda & 0 & 0 \\ 0 & 2 - \lambda & 0 \\ 0 & 0 & 3 - \lambda \end{vmatrix} = (1 - \lambda)(2 - \lambda)(3 - \lambda) = 0$$

The solutions to this cubic equation give the eigenvalues of matrix C.

1.4.5 Diagonalization of Matrices

A square matrix A is diagonalizable if it can be expressed as $A = PDP^{-1}$, where P is a matrix whose columns are eigenvectors of A, and D is a diagonal matrix whose diagonal entries are the corresponding eigenvalues.

1.4.6 Example: Diagonalization

Let $A = \begin{bmatrix} 2 & 1 \\ 1 & 3 \end{bmatrix}$. The eigenvalues and eigenvectors are calculated. If P is the matrix of eigenvectors and D is the diagonal matrix of eigenvalues, then $A = PDP^{-1}$.

1.4.7 Applications in Physics

Eigenvalues and eigenvectors play a crucial role in physics, especially in quantum mechanics. In quantum systems, eigenvalues represent measurable quantities, and eigenvectors describe the possible states of a system.

1.4.8 Eigenvalues in Dynamic Systems

In the context of dynamic systems, eigenvalues are associated with stability. For a linear system represented by matrix A, if all eigenvalues have negative real parts, the system is stable.

1.4.9   Example: Stability Analysis

Consider a dynamic system represented by $A = \begin{bmatrix} -2 & 1 \\ -1 & -3 \end{bmatrix}$. The characteristic equation is solved to find the eigenvalues. If both eigenvalues have negative real parts, the system is stable.

1.4.10   Singular Value Decomposition (SVD)

SVD is a factorization of a matrix A into three other matrices U, Σ, and V^T. It is closely related to eigendecomposition and is widely used in various applications.

1.4.11   Example: Singular Value Decomposition

Let $A = \begin{bmatrix} 1 & 2 \\ 3 & 4 \end{bmatrix}$. The singular value decomposition $A = U\Sigma V^T$ is calculated.

1.4.12   Power Iteration for Eigenvalues

Power iteration is an iterative method to find the dominant eigenvalue and corresponding eigenvector of a matrix A.

1.4.13   Example: Power Iteration

Consider a matrix $D = \begin{bmatrix} 1 & 2 \\ 3 & 4 \end{bmatrix}$. Power iteration is applied to find the dominant eigenvalue and eigenvector.

1.5   Rank of a Matrix

The rank of a matrix is a fundamental concept in linear algebra, providing insights into its properties and applications. Let's explore the definition, calculation, and significance of the rank.

1.5.1 Definition of Rank

For an $m \times n$ matrix A, the rank is the maximum number of linearly independent rows or columns in the matrix. It is denoted as rank(A).

1.5.2 Row-Reduced Echelon Form

The rank of a matrix is equal to the number of non-zero rows in its row-reduced echelon form. To find the row-reduced echelon form, Gaussian elimination or other row operations are applied.

1.5.3 Example: Calculating Rank

Consider the matrix $B = \begin{bmatrix} 1 & 2 & 3 \\ 0 & 1 & 4 \\ 2 & 3 & 5 \end{bmatrix}$. Apply row operations to find its row-reduced echelon form and determine the rank.

$$\text{Row-Reduced Echelon Form:} \quad \begin{bmatrix} 1 & 2 & 3 \\ 0 & 1 & 4 \\ 0 & 0 & 0 \end{bmatrix}$$

The rank of matrix B is 2.

1.5.4 Rank-Nullity Theorem

The rank of a matrix plus the nullity (dimension of the null space) is equal to the number of columns of the matrix. Mathematically, rank(A) + nullity(A) = n.

1.5.5 Example: Rank-Nullity Theorem

Consider a matrix $C = \begin{bmatrix} 1 & 2 & 3 \\ 2 & 4 & 6 \\ 3 & 6 & 9 \end{bmatrix}$. Find its rank and nullity. The null space can be obtained by solving the homogeneous system $Cx = 0$.

$$\text{Row-Reduced Echelon Form: } \begin{bmatrix} 1 & 2 & 3 \\ 0 & 0 & 0 \\ 0 & 0 & 0 \end{bmatrix}$$

The rank of matrix C is 1, and nullity is 2, satisfying the rank-nullity theorem.

1.5.6   Applications in Linear Systems

In the context of systems of linear equations, the rank of the coefficient matrix provides information about the solvability of the system. A system is consistent if and only if the rank of the coefficient matrix equals the rank of the augmented matrix.

1.5.7   Example: Solvability of a System

Consider a system of linear equations:

$$x + 2y + 3z = 4$$
$$2x + 4y + 6z = 8$$
$$3x + 6y + 9z = 12$$

The coefficient matrix has rank 1, and the augmented matrix also has rank 1. Therefore, the system is consistent.

1.5.8   Matrix Inversion and Full Rank

A square matrix A is invertible if and only if it is of full rank. Full rank implies that all its columns (or rows) are linearly independent.

1.5.9   Example: Matrix Inversion

Let $D = \begin{bmatrix} 1 & 2 \\ 3 & 4 \end{bmatrix}$. Determine if D is invertible by calculating its rank.

$$\text{Row-Reduced Echelon Form: } \begin{bmatrix} 1 & 2 \\ 0 & 0 \end{bmatrix}$$

The rank of matrix D is 1, indicating it is not invertible.

1.5.10 Rank in Eigenvalue Problems

In the context of eigenvalue problems, the rank of a matrix is related to the number of non-zero eigenvalues. The number of non-zero eigenvalues equals the rank of the matrix.

1.5.11 Example: Eigenvalues and Rank

Consider a matrix $E = \begin{bmatrix} 2 & 1 \\ 1 & 3 \end{bmatrix}$. Find its eigenvalues and determine the rank.

$$\text{Eigenvalues: } \lambda_1 = 4, \quad \lambda_2 = 1$$

The rank of matrix E is 2, consistent with the number of non-zero eigenvalues.

1.5.12 Singular Value Decomposition (SVD)

The rank of a matrix is also related to its singular value decomposition. The number of non-zero singular values equals the rank of the matrix.

1.5.13 Example: SVD and Rank

Consider a matrix $F = \begin{bmatrix} 1 & 0 \\ 0 & 2 \\ 0 & 0 \end{bmatrix}$. Obtain its singular value decomposition and determine the rank.

$$\text{Singular Value Decomposition: } F = U\Sigma V^T$$

The rank of matrix F is 2, consistent with the number of non-zero singular values.

1.6   System of Linear Equations

A system of linear equations is a collection of equations involving linear combinations of variables. This section explores methods for solving such systems and their applications in engineering and mathematics.

1.6.1   Definition of a System of Linear Equations

Consider a system of m linear equations with n variables:

$$a_{11}x_1 + a_{12}x_2 + \ldots + a_{1n}x_n = b_1$$

$$a_{21}x_1 + a_{22}x_2 + \ldots + a_{2n}x_n = b_2$$

$$\vdots$$

$$a_{m1}x_1 + a_{m2}x_2 + \ldots + a_{mn}x_n = b_m$$

This system can be compactly represented as $Ax = B$, where A is the coefficient matrix, x is the column vector of variables, and B is the column vector of constants.

1.6.2   Matrix Form of a System

In matrix form, the system $Ax = B$ can be expressed as:

$$\begin{bmatrix} a_{11} & a_{12} & \ldots & a_{1n} \\ a_{21} & a_{22} & \ldots & a_{2n} \\ \vdots & \vdots & \ddots & \vdots \\ a_{m1} & a_{m2} & \ldots & a_{mn} \end{bmatrix} \begin{bmatrix} x_1 \\ x_2 \\ \vdots \\ x_n \end{bmatrix} = \begin{bmatrix} b_1 \\ b_2 \\ \vdots \\ b_m \end{bmatrix}$$

1.6.3   Methods for Solving Systems

There are various methods to solve systems of linear equations, including:

Gaussian Elimination

Gaussian elimination involves transforming the augmented matrix $[A|B]$ into its row-echelon form and then back-substituting to find the solution.

Matrix Inversion

If A is invertible, the solution can be expressed as $x = A^{-1}B$.

Cramer's Rule

For a system $Ax = B$, if $\det(A) \neq 0$, the solution is given by $x_i = \frac{\det(A_i)}{\det(A)}$, where A_i is the matrix obtained by replacing the i-th column of A with vector B.

1.6.4 Example: Solving a System Using Gaussian Elimination

Consider the system:

$$2x + 3y - z = 1$$
$$4x + y + 2z = -2$$
$$3x + 2y - 3z = 3$$

Apply Gaussian elimination to find the solution.

1.6.5 Example: Solving a System Using Matrix Inversion

Consider the system:

$$x + 2y = 5$$
$$3x - y = 7$$

Express the solution using matrix inversion.

1.6.6   Example: Solving a System Using Cramer's Rule

Consider the system:

$$2x - y + z = 1$$
$$x + y + 2z = 2$$
$$3x + 2y - z = 3$$

Apply Cramer's Rule to find the solution.

1.6.7   Types of Solutions

A system of linear equations can have different types of solutions:

Unique Solution

If a system has exactly one solution, it is said to have a unique solution.

No Solution

If a system is inconsistent and has no solution, it is said to be incompatible.

Infinite Solutions

If a system has infinitely many solutions, it is called dependent or underdetermined.

1.6.8   Example: Unique Solution

Consider the system:

$$x + 2y = 3$$
$$2x + y = 5$$

Determine if the system has a unique solution.

1.6.9   Example: No Solution

Consider the system:

$$x + y = 2$$
$$x + y = 5$$

Determine if the system has a solution.

1.6.10   Example: Infinite Solutions

Consider the system:

$$x + 2y = 3$$
$$2x + 4y = 6$$

Determine if the system has infinitely many solutions.

1.6.11   Applications in Engineering

Systems of linear equations are widely used in engineering for modeling various physical phenomena, including electrical circuits, structural analysis, and fluid dynamics.

1.6.12   Example: Electrical Circuit Analysis

Consider a circuit with three resistors connected in series. The voltage across each resistor can be represented by a system of linear equations.

$$V_1 = R_1 I$$
$$V_2 = R_2 I$$
$$V_3 = R_3 I$$

Where V_1, V_2, and V_3 are the voltages, R_1, R_2, and R_3 are the resistances, and I is the current.

Chapter 2

Calculus

2.1 Functions of Single Variable

Functions of a single variable play a central role in calculus, serving as the foundation for understanding continuity, limits, derivatives, and integrals. In this section, we explore the key concepts and techniques related to functions.

2.1.1 Definition of a Function

A function f is a rule that assigns to each element x in the domain D exactly one element, denoted as $f(x)$, in the codomain E. Mathematically, $f : D \to E$.

2.1.2 Types of Functions

There are various types of functions, including:

Polynomial Functions

A polynomial function is of the form $f(x) = a_n x^n + a_{n-1} x^{n-1} + \ldots + a_1 x + a_0$, where $a_n, a_{n-1}, \ldots, a_0$ are constants.

Exponential Functions

An exponential function has the form $f(x) = a \cdot e^{bx}$, where a and b are constants and e is the base of the natural logarithm.

Trigonometric Functions

Trigonometric functions include sine (sin), cosine (cos), tangent (tan), and others. They relate the angles of a right triangle to the ratios of its sides.

2.1.3   Example: Polynomial Function

Consider the polynomial function $f(x) = 2x^3 - 5x^2 + 3x + 1$. Find its roots and analyze its behavior.

2.1.4   Example: Exponential Function

Explore the exponential function $g(x) = 3e^{2x}$. Calculate its values for different x and observe the exponential growth.

2.1.5   Limits of Functions

The limit of a function $f(x)$ as x approaches a certain value represents the behavior of f near that point. Mathematically, $\lim_{x \to a} f(x) = L$ means that $f(x)$ gets arbitrarily close to L as x gets arbitrarily close to a.

2.1.6   Example: Finding a Limit

Find the limit $\lim_{x \to 2}(x^2 + 3x - 2)$ using algebraic manipulation.

2.1.7   Continuity of Functions

A function is continuous at a point if the limit of the function at that point exists and is equal to the value of the function at that point.

2.1.8 Example: Checking Continuity

Determine the points of continuity for the function $h(x) = \frac{x+1}{x-2}$ and analyze its behavior.

2.1.9 Derivatives of Functions

The derivative of a function $f(x)$ with respect to x measures the rate at which f is changing at any given point. Mathematically, it is denoted as $f'(x)$ or $\frac{df}{dx}$.

2.1.10 Example: Finding Derivative

Find the derivative of the function $y(x) = 4x^2 - 3x + 2$ using the power rule and product rule.

2.1.11 Applications of Derivatives

Derivatives have various applications, including:

Tangent Lines

The derivative at a point gives the slope of the tangent line to the curve at that point.

Optimization

To find the maximum or minimum values of a function, set its derivative equal to zero and solve for critical points.

Rate of Change

The derivative can represent the rate of change of a quantity with respect to another.

2.1.12   Example: Tangent Line

Find the equation of the tangent line to the curve $y = x^2 - 2x + 1$ at the point $(2, 1)$.

2.1.13   Example: Optimization

Maximize the area of a rectangular garden with fixed perimeter using derivatives.

2.1.14   Integration of Functions

Integration is the reverse process of differentiation. The integral of a function $f(x)$ with respect to x is denoted as $\int f(x)\, dx$.

2.1.15   Example: Indefinite Integral

Evaluate the indefinite integral $\int (3x^2 - 2x + 1)\, dx$.

2.1.16   Example: Definite Integral

Find the area under the curve $y = \sqrt{x}$ from $x = 0$ to $x = 4$ using a definite integral.

2.1.17   Fundamental Theorem of Calculus

The Fundamental Theorem of Calculus establishes a connection between differentiation and integration. It states that if $F(x)$ is an antiderivative of $f(x)$, then $\int_a^b f(x)\, dx = F(b) - F(a)$.

2.1.18   Example: Applying Fundamental Theorem

Calculate the definite integral $\int_1^3 (2x - 1)\, dx$ using the Fundamental Theorem of Calculus.

2.1.19 Applications in Engineering

Functions of a single variable are fundamental in engineering for modeling physical phenomena, designing systems, and optimizing processes.

2.1.20 Example: Engineering Application

Model the motion of a particle using a function that describes its position over time and analyze its velocity and acceleration.

2.2 Limit, Continuity, and Differentiability

Calculus revolves around the concepts of limits, continuity, and differentiability. These concepts form the basis for understanding the behavior of functions and their derivatives. Let's delve into these fundamental aspects of calculus.

2.2.1 Limits of Functions

The limit of a function $f(x)$ as x approaches a particular value a represents the behavior of the function near a. Mathematically, $\lim_{x \to a} f(x) = L$ means that as x gets arbitrarily close to a, $f(x)$ gets arbitrarily close to L.

2.2.2 Example: Calculating a Limit

Find the limit $\lim_{x \to 2} \frac{x^2 - 4}{x - 2}$ by simplifying the expression and direct substitution.

2.2.3 Continuity of Functions

A function is continuous at a point a if the limit of the function at a exists, and the value of the function at a equals the limit. A function is continuous on an interval if it is continuous at every point within the interval.

2.2.4   Example: Checking Continuity

Examine the continuity of the function $f(x) = \sqrt{x}$ at $x = 4$ using the limit definition of continuity.

2.2.5   Differentiability of Functions

A function $f(x)$ is said to be differentiable at a point a if the derivative $f'(a)$ exists. If $f(x)$ is differentiable on an interval, it is called differentiable on that interval.

2.2.6   Example: Finding Derivative

Find the derivative of the function $g(x) = 3x^2 - 2x + 1$ using the limit definition of a derivative.

2.2.7   Limits Involving Infinity

Limits involving infinity arise when the behavior of a function becomes unbounded. Notable cases include $\lim_{x \to \infty} f(x)$ and $\lim_{x \to a} f(x) = \infty$.

2.2.8   Example: Limit at Infinity

Evaluate the limit $\lim_{x \to \infty} \frac{2x^2 + 3x - 1}{x^2 - 4}$ by dividing both numerator and denominator by x^2.

2.2.9   Continuity on Closed Intervals

A function is continuous on a closed interval $[a, b]$ if it is continuous on the open interval (a, b) and $\lim_{x \to a^+} f(x) = f(a)$ and $\lim_{x \to b^-} f(x) = f(b)$.

2.2.10   Example: Continuity on Closed Interval

Check the continuity of the function $h(x) = \frac{1}{x}$ on the closed interval $[1, 3]$ and identify any points of discontinuity.

2.2.11 Differentiability Implies Continuity

While continuity is not a sufficient condition for differentiability, differentiability implies continuity. If $f(x)$ is differentiable at a, it is continuous at a.

2.2.12 Example: Differentiability vs Continuity

Examine the function $k(x) = |x|$ and discuss its points of continuity and differentiability.

2.2.13 L'Hôpital's Rule

L'Hôpital's Rule is a powerful technique for evaluating indeterminate forms ($\frac{0}{0}$ or $\frac{\infty}{\infty}$). If $\lim_{x \to a} \frac{f(x)}{g(x)}$ is an indeterminate form, then $\lim_{x \to a} \frac{f(x)}{g(x)} = \lim_{x \to a} \frac{f'(x)}{g'(x)}$.

2.2.14 Example: Applying L'Hôpital's Rule

Evaluate the limit $\lim_{x \to 0} \frac{\sin(x)}{x}$ using L'Hôpital's Rule.

2.2.15 Continuity on Open Intervals

A function $f(x)$ can be continuous on an open interval (a, b) without being differentiable at every point in that interval.

2.2.16 Example: Continuity without Differentiability

Explore the function $m(x) = x^{\frac{2}{3}}$ and discuss its continuity and differentiability on the open interval $(-1, 1)$.

2.2.17 Intermediate Value Theorem

The Intermediate Value Theorem states that if $f(x)$ is continuous on a closed interval $[a, b]$ and k is any number between $f(a)$ and $f(b)$, then there exists at least one c in (a, b) such that $f(c) = k$.

2.2.18   Example: Applying Intermediate Value Theorem

Show that the equation $x^3 - 2x - 1 = 0$ has at least one real root in the interval $[1, 2]$ using the Intermediate Value Theorem.

2.2.19   Mean Value Theorem

The Mean Value Theorem asserts that if $f(x)$ is continuous on $[a, b]$ and differentiable on (a, b), then there exists at least one c in (a, b) such that $\frac{f(b)-f(a)}{b-a} = f'(c)$.

2.2.20   Example: Applying Mean Value Theorem

Determine the values of c guaranteed by the Mean Value Theorem for $f(x) = x^2 - 3x + 2$ on the interval $[1, 3]$.

2.2.21   Applications in Engineering

The concepts of limit, continuity, and differentiability are crucial in engineering for analyzing the behavior of physical systems, optimizing designs, and understanding rates of change.

2.2.22   Example: Engineering Application

Model the speed of a moving object using the concept of instantaneous velocity and discuss its significance in engineering analysis.

2.3   Mean Value Theorems

Mean Value Theorems are fundamental results in calculus that establish relationships between the average rate of change of a function and its instantaneous rate of change. These theorems play a key role in understanding the behavior of functions. Let's explore the main mean value theorems along with illustrative examples.

2.3.1 Rolle's Theorem

Rolle's Theorem states that if a function $f(x)$ is continuous on the closed interval $[a, b]$ and differentiable on the open interval (a, b), and $f(a) = f(b)$, then there exists at least one c in (a, b) such that $f'(c) = 0$.

2.3.2 Example: Applying Rolle's Theorem

Consider the function $h(x) = x^2 - 4x + 4$ on the interval $[1, 3]$. Verify the conditions of Rolle's Theorem and find the value of c guaranteed by the theorem.

2.3.3 Mean Value Theorem

The Mean Value Theorem (MVT) is a generalization of Rolle's Theorem. It states that if a function $f(x)$ is continuous on the closed interval $[a, b]$ and differentiable on the open interval (a, b), then there exists at least one c in (a, b) such that $f'(c) = \frac{f(b) - f(a)}{b - a}$.

2.3.4 Example: Applying Mean Value Theorem

Consider the function $g(x) = x^2$ on the interval $[1, 3]$. Apply the Mean Value Theorem to find the value of c guaranteed by the theorem.

2.3.5 Cauchy's Mean Value Theorem

Cauchy's Mean Value Theorem is a generalized form of the Mean Value Theorem. It states that if functions $f(x)$ and $g(x)$ are continuous on the closed interval $[a, b]$ and differentiable on the open interval (a, b) and $g'(x) \neq 0$ for all x in (a, b), then there exists at least one c in (a, b) such that:

$$\frac{f'(c)}{g'(c)} = \frac{f(b) - f(a)}{g(b) - g(a)}$$

2.3.6   Example: Applying Cauchy's Mean Value Theorem

Consider the functions $u(x) = e^x$ and $v(x) = x$ on the interval $[0, 1]$. Use Cauchy's Mean Value Theorem to find the value of c guaranteed by the theorem.

2.3.7   Applications in Calculus

Mean Value Theorems are valuable tools in calculus, providing insights into the behavior of functions and helping in the analysis of derivatives.

2.3.8   Example: Analyzing Functions

Apply the Mean Value Theorem to analyze the behavior of the function $f(x) = x^3 - 3x + 2$ on the interval $[-1, 1]$ and identify points guaranteed by the theorem.

2.3.9   Generalization to Higher Derivatives

Mean Value Theorems can be extended to higher derivatives. If the n-th derivative of a function exists on an interval, one can define a mean value theorem involving the n-th derivative.

2.3.10   Example: Higher Derivatives

Explore the behavior of the function $h(x) = \sin(2x)$ on the interval $[0, \frac{\pi}{2}]$ using the second derivative and the corresponding mean value theorem.

2.3.11   Cautionary Note on Assumptions

It's crucial to check the assumptions of mean value theorems before applying them. For instance, continuity and differentiability conditions must be satisfied on the specified intervals.

2.3.12   Example: Checking Assumptions

Examine the function $k(x) = \sqrt{x}$ on the interval $[0, 4]$ and verify if the assumptions of the Mean Value Theorem are met.

2.3.13   Extensions to Vector-Valued Functions

Mean Value Theorems can be generalized to vector-valued functions. For a vector-valued function $\mathbf{f}(t)$ defined on an interval $[a, b]$, there exists at least one c in (a, b) such that:

$$\mathbf{f}'(c) = \frac{\mathbf{f}(b) - \mathbf{f}(a)}{b - a}$$

2.3.14   Example: Vector-Valued Function

Consider a vector-valued function $\mathbf{r}(t) = \langle t, t^2 \rangle$ on the interval $[0, 2]$. Apply the mean value theorem for vector-valued functions.

2.4   Indeterminate Forms and L'Hôpital's Rule

Indeterminate forms arise when evaluating limits where the expression takes on an ambiguous form like $\frac{0}{0}$ or $\frac{\infty}{\infty}$. L'Hôpital's Rule provides a method to handle such cases. In this section, we explore indeterminate forms and the application of L'Hôpital's Rule.

2.4.1   Indeterminate Forms

Indeterminate forms include expressions like $\frac{0}{0}$, $\frac{\infty}{\infty}$, $0 \cdot \infty$, $\infty - \infty$, 0^0, and ∞^0. When evaluating limits involving these forms, the result is uncertain, and further analysis is required.

2.4.2   Example: Evaluating $\frac{0}{0}$

Consider the limit $\lim_{x \to 2} \frac{x^2 - 4}{x - 2}$. Simplify the expression to resolve the indeterminate form.

2.4.3   L'Hôpital's Rule

L'Hôpital's Rule provides a technique for evaluating indeterminate forms. If $\lim_{x \to a} \frac{f(x)}{g(x)}$ is an indeterminate form of $\frac{0}{0}$ or $\frac{\infty}{\infty}$, then:

$$\lim_{x \to a} \frac{f(x)}{g(x)} = \lim_{x \to a} \frac{f'(x)}{g'(x)}$$

This rule can be applied repeatedly if needed.

2.4.4   Example: Applying L'Hôpital's Rule

Evaluate the limit $\lim_{x \to 0} \frac{\sin(x)}{x}$ using L'Hôpital's Rule.

2.4.5   Extended L'Hôpital's Rule

For indeterminate forms like $0 \cdot \infty$, $\infty - \infty$, 0^0, and ∞^0, L'Hôpital's Rule can be extended. The rule states that if $\lim_{x \to a}[f(x) \cdot g(x)]$ is an indeterminate form, then:

$$\lim_{x \to a}[f(x) \cdot g(x)] = \lim_{x \to a} \frac{f(x)}{\frac{1}{g(x)}}$$

2.4.6   Example: Applying Extended L'Hôpital's Rule

Evaluate the limit $\lim_{x \to 0} x \cdot \ln(x)$ using the extended form of L'Hôpital's Rule.

2.4.7   Handling $\infty - \infty$ Forms

If $\lim_{x \to a}[f(x) - g(x)]$ is of the form $\infty - \infty$, the expression can be rewritten to utilize L'Hôpital's Rule.

2.4.8   Example: Handling $\infty - \infty$

Evaluate the limit $\lim_{x \to 1} \frac{x^2 - 1}{\ln(x)}$ by transforming it into the $\infty - \infty$ form.

Indeterminate Forms Involving 0^0 and 1^∞

When dealing with indeterminate forms like 0^0 or 1^∞, logarithmic manipulation can be applied before applying L'Hôpital's Rule.

Example: Handling 0^0 Form

Evaluate the limit $\lim_{x \to 0^+} x^x$ by transforming it into a form suitable for L'Hôpital's Rule.

Example: Handling 1^∞ Form

Evaluate the limit $\lim_{x \to \infty} \left(1 + \frac{1}{x}\right)^x$ by transforming it into a form suitable for L'Hôpital's Rule.

2.4.9   Indeterminate Forms and Trigonometric Limits

Indeterminate forms involving trigonometric functions can be resolved by applying trigonometric identities or L'Hôpital's Rule.

2.4.10   Example: Trigonometric Limit

Evaluate the limit $\lim_{x \to 0} \frac{\tan(x)}{x}$ by applying L'Hôpital's Rule.

2.4.11   Application to Improper Integrals

L'Hôpital's Rule is also applicable to evaluate certain types of improper integrals.

2.4.12   Example: Improper Integral

Evaluate the improper integral $\int_0^1 \frac{\ln(x)}{1-x}\,dx$ using L'Hôpital's Rule.

2.4.13   Cautionary Note on Applying L'Hôpital's Rule

While L'Hôpital's Rule is a powerful tool, it should be applied judiciously. Care must be taken to ensure that the assumptions of the rule are met.

2.4.14   Example: Misapplication of L'Hôpital's Rule

Examine the limit $\lim_{x \to 0} \frac{\sin(x)}{x^2}$ and discuss the potential misapplication of L'Hôpital's Rule.

2.4.15   Applications in Engineering

L'Hôpital's Rule finds applications in various engineering scenarios, particularly in analyzing limits and solving mathematical problems arising in engineering problems.

2.4.16   Example: Engineering Application

Apply L'Hôpital's Rule to analyze the behavior of a function representing the speed of a moving object and discuss its implications in engineering.

2.5   Taylor's Theorem

Taylor's Theorem is a powerful tool in calculus that allows us to approximate a function using its derivatives. It provides a way to express a function as an infinite series, known as the Taylor series. Let's delve into the details of Taylor's Theorem and its applications.

2.5.1   Taylor's Theorem Statement

Let $f(x)$ be a function with $n+1$ continuous derivatives on an open interval containing a. Then, Taylor's Theorem states that for any x in that interval, there exists a c between a and x such that:

$$f(x) = f(a) + f'(a)(x-a) + \frac{f''(a)}{2!}(x-a)^2 + \cdots + \frac{f^{(n)}(a)}{n!}(x-a)^n + R_n(x)$$

Here, $R_n(x)$ is the remainder term, given by:

$$R_n(x) = \frac{f^{(n+1)}(c)}{(n+1)!}(x-a)^{n+1}$$

2.5.2   Taylor Series

The Taylor series for a function $f(x)$ centered at a is the infinite sum:

$$f(x) = f(a) + f'(a)(x-a) + \frac{f''(a)}{2!}(x-a)^2 + \frac{f^{(3)}(a)}{3!}(x-a)^3 + \cdots$$

Example: Taylor Series for e^x

Find the Taylor series for e^x centered at $a = 0$ and express it in sigma notation.

2.5.3   Maclaurin Series

If $a = 0$, the Taylor series is called the Maclaurin series. It simplifies to:

$$f(x) = f(0) + f'(0)x + \frac{f''(0)}{2!}x^2 + \frac{f^{(3)}(0)}{3!}x^3 + \cdots$$

2.5.4   Example: Maclaurin Series for $\sin(x)$

Find the Maclaurin series for $\sin(x)$ and write it in sigma notation.

2.5.5   Error in Taylor Approximation

The remainder term $R_n(x)$ in Taylor's Theorem represents the error in the approximation. The larger n, the smaller the error. The formula allows us to estimate how accurate our approximation is.

2.5.6   Example: Estimating Error

Estimate the error when approximating e using the first three terms of the Taylor series for e^x centered at $a = 0$.

2.5.7   Convergence of Taylor Series

The Taylor series converges to the function if the limit of the remainder term approaches zero as n approaches infinity. This is known as Taylor's Remainder Theorem.

2.5.8   Example: Testing Convergence

Examine the convergence of the Taylor series for e^x centered at $a = 0$ by applying Taylor's Remainder Theorem.

2.5.9   Interval of Convergence

The interval of convergence of a Taylor series is the range of x values for which the series converges to the function. It can be determined using the ratio test.

2.5.10   Example: Determining Interval of Convergence

Determine the interval of convergence for the Taylor series of $\ln(1 + x)$ centered at $a = 0$ using the ratio test.

2.5.11   Multivariable Taylor Series

Taylor's Theorem can be extended to functions of multiple variables. The multivariable Taylor series approximates a function using partial derivatives.

2.5.12   Example: Multivariable Taylor Series

Find the second-degree Taylor polynomial for the function $f(x, y) = e^{xy}$ centered at $(a, b) = (0, 0)$.

2.5.13   Applications in Engineering

Taylor's Theorem is extensively used in engineering for approximation and error analysis. It plays a crucial role in fields such as control systems, signal processing, and numerical methods.

2.5.14   Example: Engineering Application

Apply Taylor's Theorem to approximate the behavior of a dynamic system described by a nonlinear function in the vicinity of an equilibrium point.

2.5.15 Taylor Series for Common Functions

Knowing the Taylor series for common functions simplifies complex calculations. Common Taylor series include those for exponential, trigonometric, and logarithmic functions.

2.5.16 Example: Using Taylor Series

Express the function $f(x) = \cos^2(x)$ as a Taylor series centered at $a = 0$ up to the fourth degree.

2.6 Maxima and Minima

In calculus, finding the maximum and minimum values of a function is crucial for optimization problems and understanding the behavior of a system. The process involves analyzing critical points, points of inflection, and endpoints. Let's explore the concepts and methods associated with maxima and minima.

2.6.1 Local Extrema

A function $f(x)$ has a local maximum at a point c if $f(c)$ is greater than or equal to $f(x)$ for all x in some open interval containing c. Similarly, $f(x)$ has a local minimum at c if $f(c)$ is less than or equal to $f(x)$ for all x in some open interval containing c.

2.6.2 Critical Points

Critical points are values of x where $f'(x) = 0$ or $f'(x)$ does not exist. These points are potential locations of maxima or minima.

2.6.3 Example: Finding Critical Points

Consider the function $f(x) = x^3 - 3x^2 + 2x$. Find the critical points and determine whether they correspond to maxima or minima.

2.6.4   First Derivative Test

The First Derivative Test is a method to determine whether a critical point corresponds to a local maximum or minimum. If $f'(x)$ changes from positive to negative at a critical point c, then f has a local maximum at c. If $f'(x)$ changes from negative to positive, then f has a local minimum at c.

2.6.5   Example: Applying First Derivative Test

Apply the First Derivative Test to analyze the critical points of the function $g(x) = x^4 - 4x^3 + 6x^2$.

2.6.6   Second Derivative Test

The Second Derivative Test is another criterion for determining the nature of critical points. If $f'(c) = 0$ and $f''(c) > 0$, then f has a local minimum at c. If $f'(c) = 0$ and $f''(c) < 0$, then f has a local maximum at c.

2.6.7   Example: Applying Second Derivative Test

Consider the function $h(x) = 3x^4 - 16x^3 + 18x^2$. Use the Second Derivative Test to analyze the critical points.

2.6.8   Absolute Extrema

Absolute maxima and minima are the highest and lowest points on the entire domain of a function. These can occur at critical points, endpoints, or points of discontinuity.

2.6.9   Example: Finding Absolute Extrema

Find the absolute maximum and minimum values of the function $f(x) = 2x^3 - 3x^2 - 12x + 5$ on the interval $[-2, 3]$.

2.6.10 Global Optimization

Global optimization involves finding the absolute maximum or minimum value of a function over its entire domain. This often requires a combination of critical point analysis and endpoint considerations.

2.6.11 Example: Global Optimization

Consider the function $p(x) = x^3 - 6x^2 + 9x + 2$. Determine the global maximum and minimum values over the interval $[-1, 4]$.

2.6.12 Applications in Engineering Design

Finding maxima and minima is fundamental in engineering design, where engineers strive to optimize various parameters, such as cost, efficiency, or performance.

2.6.13 Example: Engineering Optimization

An engineer designing a rectangular enclosure wants to minimize the cost of fencing. Apply calculus to find the dimensions that result in the minimum cost.

2.6.14 Lagrange Multipliers

Lagrange Multipliers are a method for finding maxima and minima of a function subject to equality constraints. This technique is particularly useful in optimization problems with constraints.

2.6.15 Example: Lagrange Multipliers

Optimize the function $f(x, y) = x^2 + y^2$ subject to the constraint $x + y = 10$ using Lagrange Multipliers.

2.6.16   Sensitivity Analysis

Sensitivity analysis involves assessing how changes in parameters or constraints affect the optimal solution. It is crucial in engineering and economics.

2.6.17   Example: Sensitivity Analysis

Analyze the sensitivity of the optimal solution in a manufacturing process optimization problem when there are variations in raw material costs.

2.7   Integration and its Applications

Integration is a fundamental concept in calculus that involves finding the antiderivative of a function. It has various applications in engineering, physics, and other sciences. In this section, we explore integration techniques and their practical applications.

2.7.1   Definite and Indefinite Integrals

The definite integral represents the signed area under a curve, while the indefinite integral represents a family of antiderivatives. The fundamental theorem of calculus establishes a connection between these two concepts.

2.7.2   Example: Evaluating a Definite Integral

Compute the definite integral $\int_0^2 (2x - 1)\, dx$ to find the area under the curve.

2.7.3   Integration Techniques

Several techniques, such as substitution, integration by parts, and partial fractions, are used to evaluate integrals. These methods are valuable for handling complex integrands.

2.7.4 Example: Substitution Method

Evaluate the integral $\int e^{2x} \sin(3x)\, dx$ using the substitution $u = e^{2x}$.

2.7.5 Example: Integration by Parts

Apply integration by parts to evaluate the integral $\int x \cos(x)\, dx$.

2.7.6 Applications in Physics: Work

Integration is extensively used in physics to calculate work done by a force. The work done is given by the integral of the force with respect to displacement.

2.7.7 Example: Calculating Work

Determine the work done by a force $F(x) = 3x^2$ in moving an object from $x = 1$ to $x = 4$.

2.7.8 Applications in Engineering: Fluid Pressure

Integration is crucial in determining fluid pressure. The pressure at a certain depth in a fluid is given by the integration of the fluid density and gravitational acceleration.

2.7.9 Example: Fluid Pressure Calculation

Calculate the pressure at a depth of 10 meters in a fluid with a density of 1000 kg/m^3.

2.7.10 Applications in Economics: Revenue

In economics, integration is used to calculate revenue. The revenue generated by selling a certain quantity of a product is given by the integral of the demand function.

2.7.11   Example: Revenue Calculation

Determine the revenue function for a product with a demand function $p(x) = 20 - x$ and find the total revenue when selling from $x = 1$ to $x = 5$.

2.7.12   Improper Integrals

Improper integrals involve integrating over unbounded intervals or integrating functions with infinite discontinuities.

2.7.13   Example: Evaluating an Improper Integral

Evaluate the improper integral $\int_1^\infty \frac{1}{x^2}\, dx$ to determine if it converges or diverges.

2.7.14   Applications in Probability: Cumulative Distribution Function

In probability theory, integration is used to calculate cumulative distribution functions (CDFs) that describe the probability of a random variable being less than or equal to a specific value.

2.7.15   Example: Probability Calculation

Find the cumulative distribution function for a continuous random variable with a given probability density function.

2.7.16   Applications in Signal Processing: Fourier Transform

The Fourier transform, a type of integral transform, is widely used in signal processing. It represents a function as a sum of sinusoidal functions.

2.7.17 Example: Fourier Transform

Apply the Fourier transform to find the frequency spectrum of a signal defined over a specific time interval.

2.7.18 Applications in Geometry: Surface Area

Integration is employed to calculate the surface area of three-dimensional objects. The surface area is given by integrating the element of surface area over the object.

2.7.19 Example: Surface Area Calculation

Determine the surface area of a cone with a given radius and height.

2.8 Differential Equations

Differential equations are fundamental in expressing relationships involving rates of change. They find wide applications in engineering, physics, biology, and various other fields. In this section, we explore different types of differential equations and methods to solve them.

2.8.1 First-Order Ordinary Differential Equations

First-order ordinary differential equations involve only the first derivative of the unknown function. They can be solved using various methods, including separation of variables and integrating factors.

2.8.2 Example: Separation of Variables

Consider the differential equation $\frac{dy}{dx} = 2x$. Apply separation of variables to find the solution.

2.8.3   Example: Integrating Factor

Solve the first-order differential equation $\frac{dy}{dx} + 3y = 6$ using the integrating factor method.

2.8.4   Second-Order Ordinary Differential Equations

Second-order ordinary differential equations involve the second derivative of the unknown function. Common methods for solving them include the characteristic equation and undetermined coefficients.

2.8.5   Example: Characteristic Equation

Solve the second-order differential equation $y'' - 4y = 0$ using the characteristic equation.

2.8.6   Example: Undetermined Coefficients

Find the solution to the differential equation $y'' + y = \sin(x)$ using the undetermined coefficients method.

2.8.7   Systems of Differential Equations

In engineering, many problems involve multiple variables evolving over time. Systems of differential equations provide a framework for modeling and solving such scenarios.

2.8.8   Example: Linear System

Explore a system of linear differential equations representing the dynamics of interconnected mechanical components.

2.8.9   Partial Differential Equations

Partial differential equations involve partial derivatives with respect to multiple independent variables. They are prevalent in physics and engineering, describing phenomena like heat conduction and wave propagation.

2.8.10   Example: Heat Equation

Examine the one-dimensional heat equation $\frac{\partial u}{\partial t} = k \frac{\partial^2 u}{\partial x^2}$ and its solution.

2.8.11   Example: Wave Equation

Study the one-dimensional wave equation $\frac{\partial^2 u}{\partial t^2} = c^2 \frac{\partial^2 u}{\partial x^2}$ and its application to wave propagation.

2.8.12   Numerical Methods for Differential Equations

In many practical situations, analytical solutions to differential equations are challenging to obtain. Numerical methods, such as Euler's method and the Runge-Kutta method, provide approximate solutions.

2.8.13   Example: Euler's Method

Apply Euler's method to approximate the solution of the differential equation $\frac{dy}{dx} = x^2 - y^2$ with an initial condition.

2.8.14   Example: Runge-Kutta Method

Utilize the Runge-Kutta method to numerically solve a system of coupled differential equations describing population dynamics.

2.8.15   Boundary Value Problems

Boundary value problems involve finding a solution that satisfies certain conditions at the boundaries of the domain. These problems commonly arise in physics and engineering.

2.8.16   Example: Sturm-Liouville Problem

Explore the Sturm-Liouville problem $\frac{d}{dx}\left(p(x)\frac{dy}{dx}\right) + q(x)y = \lambda r(x)y$ and its application to vibrating strings.

2.8.17   Applications in Engineering: Control Systems

Differential equations play a crucial role in modeling and analyzing dynamic systems, particularly in control engineering. They describe the time evolution of system variables.

2.8.18   Example: Control System Dynamics

Model the dynamics of a simple control system using differential equations and analyze the system's response to different inputs.

2.8.19   Applications in Physics: Newton's Law of Cooling

Newton's Law of Cooling, describing the rate of temperature change in an object, is modeled using a first-order ordinary differential equation.

2.8.20   Example: Newton's Law of Cooling

Apply Newton's Law of Cooling to determine the temperature of a hot beverage as it cools in a room with a constant temperature.

Chapter 3

Differential Equations

3.1 First-order linear and nonlinear differential equations

Differential equations are mathematical expressions that involve derivatives and are fundamental in modeling dynamic systems. First-order differential equations play a crucial role in various engineering applications. In this section, we will explore both linear and nonlinear first-order differential equations and their solutions.

3.1.1 First-order Linear Differential Equations

A first-order linear differential equation has the form:

$$\frac{dy}{dx} + P(x)y = Q(x)$$

where $P(x)$ and $Q(x)$ are functions of x. The solution can be found using an integrating factor, typically denoted as $e^{\int P(x)\,dx}$.

Example: Solving Linear Differential Equation

Consider the differential equation $\frac{dy}{dx} + 2xy = x^2$. Find the solution using an integrating factor.

Numerical Example: Linear Differential Equation

Utilize a numerical method, such as Euler's method, to approximate the solution of the linear differential equation with an initial condition.

3.1.2   First-order Nonlinear Differential Equations

Nonlinear first-order differential equations are equations where the dependent variable and its derivative appear nonlinearly. These equations often require specific techniques for solving.

Example: Separable Differential Equation

Explore the separable differential equation $y' = xy^2$. Find the solution by separating variables and integrating.

Example: Exact Differential Equation

Consider the exact differential equation $y' = 2xy + x^2$. Determine the solution using the method of exact differentials.

Numerical Example: Nonlinear Differential Equation

Apply a numerical method, such as the Runge-Kutta method, to solve a nonlinear first-order differential equation with given initial conditions.

3.1.3   Applications in Engineering: RC Circuits

First-order linear differential equations are prevalent in electrical engineering, particularly in modeling RC circuits. The voltage across a capacitor in an RC circuit can be described using a first-order linear differential equation.

Example: RC Circuit Differential Equation

Derive the first-order linear differential equation governing the voltage across a capacitor in an RC circuit and find its solution.

3.1.4 Applications in Biology: Population Growth

Nonlinear differential equations find applications in biology, such as modeling population growth. The logistic equation is a common example.

Example: Logistic Equation

Explore the logistic equation $y' = ky(1 - y)$ representing population growth, where k is a positive constant.

3.1.5 Applications in Chemistry: Reaction Kinetics

First-order differential equations are used in chemistry to describe reaction kinetics. The rate of a chemical reaction can be modeled using a first-order linear differential equation.

Example: Chemical Kinetics

Formulate a first-order linear differential equation to model the rate of decay of a reactant in a chemical reaction and find its solution.

3.1.6 Applications in Economics: Exponential Growth

Economics often involves modeling scenarios with exponential growth or decay. First-order differential equations are employed to describe such economic phenomena.

Example: Economic Growth Model

Develop a first-order differential equation to model exponential economic growth and analyze its behavior over time.

3.1.7   Systems of First-order Differential Equations

In engineering, systems of first-order differential equations are used to model interconnected dynamic systems. These systems can be solved using matrix methods or numerical techniques.

Example: Mechanical System

Explore a system of first-order linear differential equations representing the dynamics of interconnected mechanical components.

Chapter 4

Differential Equations

4.1 Higher-order linear differential equations with constant coefficients

Higher-order linear differential equations with constant coefficients are a class of differential equations frequently encountered in engineering and physics. They are characterized by having derivatives of the dependent variable up to a certain order, multiplied by constant coefficients. In this section, we will delve into the theory and solution methods for these equations.

4.1.1 General Form of the Equation

A general nth-order linear differential equation with constant coefficients can be expressed as:

$$a_n y^{(n)} + a_{n-1} y^{(n-1)} + \ldots + a_1 y' + a_0 y = F(x)$$

where $a_n, a_{n-1}, \ldots, a_1, a_0$ are constants, y is the dependent variable, $F(x)$ is the non-homogeneous term, and $y^{(k)}$ denotes the kth derivative of y with respect to x.

Characteristic Equation

The solutions to homogeneous linear differential equations can be found by solving the characteristic equation:

$$a_n r^n + a_{n-1} r^{n-1} + \ldots + a_1 r + a_0 = 0$$

The roots of this equation $(r_1, r_2, \ldots, r_n)$ determine the form of the homogeneous solution.

Example: Solving Homogeneous Equation

Consider the equation $y'' - 3y' + 2y = 0$. Find the homogeneous solution using the characteristic equation.

Particular Solution

The particular solution (y_p) for non-homogeneous linear differential equations can be found based on the form of $F(x)$. It is added to the homogeneous solution to obtain the general solution.

Example: Solving Non-homogeneous Equation

For the equation $y'' - 3y' + 2y = e^{2x}$, find the particular solution using the method of undetermined coefficients.

Complex Roots of the Characteristic Equation

When the characteristic equation has complex roots, the homogeneous solution involves complex exponentials. Euler's formula is often employed in such cases.

Example: Complex Roots

Explore the equation $y'' + 4y = 0$, which has complex roots. Find the homogeneous solution using Euler's formula.

4.1.2 Applications in Mechanical Engineering: Vibrations

Higher-order linear differential equations are commonly used to model vibrations in mechanical systems. The equation for simple harmonic motion is a second-order linear differential equation.

Example: Simple Harmonic Motion

Derive the differential equation governing simple harmonic motion and analyze its solutions for different initial conditions.

4.1.3 Applications in Electrical Engineering: LRC Circuits

LRC circuits, consisting of inductors (L), resistors (R), and capacitors (C), are modeled using higher-order linear differential equations.

Example: LRC Circuit Equation

Formulate the differential equation for an LRC circuit and determine the conditions for underdamping, overdamping, and critical damping.

4.1.4 Applications in Physics: Newton's Law of Motion

Newton's second law of motion, $F = ma$, can be expressed as a second-order linear differential equation for displacement.

Example: Mass-Spring System

Apply Newton's law to derive the differential equation for a mass-spring system and investigate its solutions.

4.1.5 Numerical Solutions

In some cases, analytical solutions may be challenging to obtain. Numerical methods, such as the Runge-Kutta method, can be employed to approximate solutions.

Example: Runge-Kutta Method

Use the Runge-Kutta method to approximate the solution of a third-order linear differential equation with constant coefficients.

4.2 Cauchy's and Euler's Equations

Cauchy's and Euler's equations are differential equations that find applications in various branches of engineering and physics. They are fundamental in solving linear differential equations with variable coefficients. In this section, we will explore these equations, their solutions, and their significance in different contexts.

4.2.1 Cauchy's Equation

Cauchy's equation is a second-order linear differential equation with variable coefficients. It is given by:

$$a(x)y'' + b(x)y' + c(x)y = 0$$

where $a(x), b(x)$, and $c(x)$ are functions of x. The solutions to Cauchy's equation often involve special functions, such as Bessel functions.

Example: Cauchy's Equation Solution

Consider the Cauchy's equation $x^2 y'' - xy' + (x^2 - \frac{1}{4})y = 0$. Find the solution using power series and Bessel functions.

Numerical Example: Cauchy's Equation

Use numerical methods, like the shooting method, to approximate the solution of Cauchy's equation with given boundary conditions.

4.2.2   Euler's Equation

Euler's equation is a second-order linear differential equation with variable co-efficients. It is given by:

$$x^2 y'' + pxy' + qy = 0$$

where p and q are constants. Euler's equation has solutions that involve power series, and it often arises in problems with cylindrical symmetry.

Example: Euler's Equation Solution

Explore the Euler's equation $x^2 y'' + 2xy' + 2y = 0$ and find the solution using power series and transformation methods.

Applications in Engineering: Cylindrical Symmetry

Euler's equation is commonly encountered in engineering problems with cylindrical symmetry, such as heat conduction in a cylindrical rod.

Example: Heat Conduction in a Rod

Model the heat conduction in a cylindrical rod using Euler's equation and determine the temperature distribution.

4.2.3   Cauchy-Euler Equation

The Cauchy-Euler equation is a special case that combines features of both Cauchy's and Euler's equations. It is given by:

$$ax^2 y'' + bxy' + cy = 0$$

where a, b, and c are constants. The solutions to the Cauchy-Euler equation involve power series and have applications in problems with radial symmetry.

Example: Cauchy-Euler Equation Solution

Consider the Cauchy-Euler equation $x^2 y'' - xy' + y = 0$ and find the solution using power series and transformation methods.

Applications in Physics: Quantum Mechanics

The Cauchy-Euler equation appears in the radial part of the Schrödinger equation in quantum mechanics, describing the behavior of electrons in atoms.

Example: Schrödinger Equation

Explore the radial part of the Schrödinger equation for a hydrogen atom, which involves the Cauchy-Euler equation, and analyze its solutions.

4.3 Laplace Transforms

Laplace transforms are powerful mathematical tools used to solve linear differential equations with constant coefficients. They provide a convenient way to transform differential equations into algebraic equations, making the solution process more straightforward. In this section, we will explore the Laplace transform, its properties, and its applications in solving differential equations.

4.3.1 Definition of Laplace Transform

The Laplace transform of a function $f(t)$ is defined as:

$$\mathcal{L}\{f(t)\} = F(s) = \int_0^\infty e^{-st} f(t)\, dt$$

where s is a complex number. The Laplace transform converts a function of time t into a function of the complex variable s.

4.3.2 Properties of Laplace Transform

Laplace transforms have several useful properties that facilitate the solution of differential equations. Some key properties include linearity, the Laplace transform of derivatives, and the shifting property.

Linearity

The Laplace transform is a linear operation, meaning that for constants a and b and functions $f(t)$ and $g(t)$, we have:

$$\mathcal{L}\{af(t) + bg(t)\} = a\mathcal{L}\{f(t)\} + b\mathcal{L}\{g(t)\}$$

Derivative Property

The Laplace transform of the derivative of a function $f(t)$ is given by:

$$\mathcal{L}\{f'(t)\} = s\mathcal{L}\{f(t)\} - f(0)$$

Shifting Property

If $F(s) = \mathcal{L}\{f(t)\}$, then the Laplace transform of $e^{at}f(t)$ is given by:

$$\mathcal{L}\{e^{at}f(t)\} = F(s - a)$$

4.3.3 Inverse Laplace Transform

The inverse Laplace transform allows us to recover the original function $f(t)$ from its Laplace transform $F(s)$. The inverse Laplace transform is denoted as $\mathcal{L}^{-1}$.

$$\mathcal{L}^{-1}\{F(s)\} = f(t)$$

4.3.4   Solving Differential Equations

Laplace transforms are particularly useful in solving linear constant-coefficient differential equations. The Laplace transform of the differential equation transforms it into an algebraic equation, making it easier to solve for the unknown function.

Example: Solving First-Order Differential Equation

Consider the first-order linear differential equation $y'(t)+2y(t) = 3$. Use Laplace transforms to find the solution.

Example: Solving Second-Order Differential Equation

Explore the solution to the second-order linear differential equation $y''(t) + 4y(t) = \sin(2t)$ using Laplace transforms.

Numerical Example: Laplace Transform and Differential Equations

Use Laplace transforms to solve a system of coupled differential equations numerically, incorporating initial conditions.

4.3.5   Applications in Control Systems: Transfer Functions

Laplace transforms are widely used in control systems engineering to analyze and design dynamic systems. The transfer function, which relates the Laplace transform of the output to the input, is a key concept.

Example: Transfer Function

Derive the transfer function of a simple control system and analyze its behavior in the Laplace domain.

4.3.6 Applications in Electrical Circuits: Response Analysis

Laplace transforms are employed in electrical engineering to analyze the transient and steady-state response of circuits to different inputs.

Example: Circuit Response

Analyze the response of an electrical circuit to a step input using Laplace transforms and determine key performance parameters.

4.4 Partial Differential Equations and Their Solutions

Partial Differential Equations (PDEs) play a crucial role in modeling physical phenomena with multiple independent variables. They arise in various engineering applications, describing how quantities change in space and time. In this section, we will explore different types of PDEs, solution methods, and applications.

4.4.1 Classification of PDEs

PDEs are classified based on their order and linearity. The order of a PDE is determined by the highest-order derivative present. Linear PDEs have solutions that can be expressed as a linear combination of functions and their derivatives.

First-Order Linear PDE

An example of a first-order linear PDE is:

$$a\frac{\partial u}{\partial x} + b\frac{\partial u}{\partial y} = c$$

Second-Order Linear PDE

A common second-order linear PDE is the heat equation:

$$\frac{\partial u}{\partial t} = \alpha \left(\frac{\partial^2 u}{\partial x^2} + \frac{\partial^2 u}{\partial y^2} \right)$$

Nonlinear PDE

An example of a nonlinear PDE is the Burgers' equation:

$$\frac{\partial u}{\partial t} + u \frac{\partial u}{\partial x} = \nu \frac{\partial^2 u}{\partial x^2}$$

4.4.2 Method of Separation of Variables

The method of separation of variables is a common technique to solve linear PDEs. It involves assuming a solution of the form $u(x, y, t) = X(x)Y(y)T(t)$ and substituting it into the PDE.

Example: Heat Equation

Consider the one-dimensional heat equation $\frac{\partial u}{\partial t} = \alpha \frac{\partial^2 u}{\partial x^2}$. Use the method of separation of variables to find the solution.

Numerical Example: Finite Difference Method

Apply the finite difference method to solve a two-dimensional heat equation numerically, considering a rectangular domain.

4.4.3 Characteristic Curves for Nonlinear PDEs

Nonlinear PDEs often involve characteristic curves, along which the solution satisfies ordinary differential equations (ODEs). The method of characteristics is employed to find solutions.

Example: Burgers' Equation

Explore the characteristics of Burgers' equation $\frac{\partial u}{\partial t} + u\frac{\partial u}{\partial x} = \nu\frac{\partial^2 u}{\partial x^2}$ and find the solution.

4.4.4 Applications in Fluid Dynamics: Navier-Stokes Equation

The Navier-Stokes equation is a set of nonlinear PDEs governing fluid flow. It describes the conservation of momentum for incompressible fluid.

Example: Two-Dimensional Flow

Investigate the two-dimensional Navier-Stokes equation for incompressible flow and analyze the solutions in specific scenarios.

4.4.5 Applications in Structural Mechanics: Wave Equation

The wave equation is a second-order linear PDE that describes the propagation of waves, such as vibrations in structures.

Example: Vibrations in a String

Apply the wave equation to model vibrations in a stretched string and determine the characteristics of the resulting waves.

Chapter 5

Complex Analysis

5.1 Analytic Functions

Analytic functions, also known as holomorphic functions, play a central role in complex analysis. These functions have well-defined derivatives at every point within their domain. In this section, we will explore the properties of analytic functions, understand the Cauchy-Riemann equations, and examine their applications.

5.1.1 Definition of Analytic Functions

A complex function $f(z)$ is said to be analytic in a domain D if it is differentiable at every point z within D. The derivative $f'(z)$ exists, and this property holds in an open set containing D.

Cauchy-Riemann Equations

For a function $f(z) = u(x,y) + iv(x,y)$ to be analytic, it must satisfy the Cauchy-Riemann equations:

$$\frac{\partial u}{\partial x} = \frac{\partial v}{\partial y} \quad \text{and} \quad \frac{\partial u}{\partial y} = -\frac{\partial v}{\partial x}$$

63

These equations ensure that the real and imaginary parts of $f(z)$ satisfy the conditions for differentiability.

Example: Verifying Analyticity

Consider the function $f(z) = e^z = e^{x+iy}$. Verify its analyticity by checking the Cauchy-Riemann equations.

5.1.2   Properties of Analytic Functions

Analytic functions exhibit several important properties, including the preservation of angles and conformal mapping.

Angle Preservation

Analytic functions preserve angles between curves in their domain. This property is a consequence of the Cauchy-Riemann equations.

Conformal Mapping

Analytic functions define conformal mappings, meaning that they locally preserve angles and magnitudes. This property is valuable in various engineering applications, such as heat conduction problems.

5.1.3   Complex Integration and Cauchy's Theorem

Analytic functions are particularly significant in complex integration, leading to Cauchy's theorem. This theorem states that the integral of an analytic function over a closed contour is zero.

Example: Cauchy's Integral Formula

Utilize Cauchy's integral formula to evaluate the integral of an analytic function over a closed contour.

Applications in Engineering: Electrostatics

In electrostatics, the electric potential is described by an analytic function. Understanding the properties of analytic functions is crucial for analyzing electric fields.

Example: Electric Field Potential

Model the electric potential using an analytic function and analyze its behavior to determine the electric field.

5.1.4 Singularities and Residues

Analytic functions may have singularities, points where they are not defined or not analytic. Residues are important in evaluating integrals involving singularities.

Classification of Singularities

Singularities can be classified as removable, poles, or essential. Each type has distinct characteristics.

Example: Residue Calculation

Compute the residue of an analytic function at a given singularity to evaluate a complex integral.

5.1.5 Applications in Signal Processing: Laplace Transform

Analytic functions play a role in signal processing through the Laplace transform. Understanding their properties aids in analyzing signals.

Example: Laplace Transform

Apply the Laplace transform to a time-domain signal represented by an analytic function.

5.2   Cauchy-Riemann Equations

The Cauchy-Riemann equations are a set of partial differential equations that characterize the conditions for a complex function to be analytic. These equations are central to complex analysis and provide a fundamental criterion for the differentiability of a complex function.

5.2.1   Derivation of Cauchy-Riemann Equations

Consider a complex function $f(z) = u(x, y) + iv(x, y)$, where $z = x + iy$. The Cauchy-Riemann equations are derived by equating the real and imaginary parts of the complex derivative:

$$\frac{\partial u}{\partial x} = \frac{\partial v}{\partial y} \quad \text{and} \quad \frac{\partial u}{\partial y} = -\frac{\partial v}{\partial x}$$

These equations express the relationship between the partial derivatives of the real and imaginary parts.

Example: Verifying Cauchy-Riemann Equations

Consider the complex function $f(z) = z^2 = (x + iy)^2$. Verify the Cauchy-Riemann equations for this function.

5.2.2   Interpretation of Cauchy-Riemann Equations

The Cauchy-Riemann equations are essential for understanding the geometric interpretation of analyticity. They ensure that the real and imaginary parts of an analytic function have continuous first-order partial derivatives and satisfy a specific relationship.

Geometric Interpretation

The Cauchy-Riemann equations guarantee that the vector field formed by the gradient of the real part is orthogonal to the vector field formed by the gradient of the imaginary part.

Example: Geometric Interpretation

Illustrate the geometric interpretation of the Cauchy-Riemann equations using a specific analytic function and visualize the orthogonality of the vector fields.

5.2.3 Polar Form and Cauchy-Riemann Equations

Expressing complex numbers in polar form ($z = re^{i\theta}$) provides an alternative perspective on the Cauchy-Riemann equations.

Polar Form of Cauchy-Riemann Equations

The Cauchy-Riemann equations in polar form are given by:

$$\frac{\partial u}{\partial r} = \frac{1}{r}\frac{\partial v}{\partial \theta} \quad \text{and} \quad \frac{1}{r}\frac{\partial u}{\partial \theta} = -\frac{\partial v}{\partial r}$$

These equations highlight the connection between the radial and angular derivatives.

Example: Polar Form Analysis

Analyze a complex function in polar form, applying the Cauchy-Riemann equations to explore the relationship between the radial and angular derivatives.

5.2.4 Applications in Fluid Dynamics: Complex Potential

In fluid dynamics, the concept of a complex potential is employed to describe irrotational flow. The Cauchy-Riemann equations play a crucial role in this context.

Complex Potential Formulation

Express the complex potential for irrotational flow and demonstrate how the Cauchy-Riemann equations are satisfied.

Example: Irrotational Flow

Apply the concept of complex potential to model irrotational flow around a circular object and analyze the flow pattern.

5.2.5   Numerical Examples: Solving Equations

Utilize numerical methods, such as finite differences or finite elements, to solve equations derived from the Cauchy-Riemann equations in specific scenarios.

Numerical Solution: Circular Domain

Numerically solve the Cauchy-Riemann equations in a circular domain, demonstrating how numerical methods can be applied to analyze complex functions.

5.3   Contour Integration

Contour integration is a powerful technique in complex analysis used to evaluate complex integrals along curves in the complex plane. This section explores the fundamental concepts of contour integration, introduces the Cauchy's Integral Formula, and discusses residues.

5.3.1   Introduction to Contour Integration

Contour integration involves integrating complex functions along curves or contours in the complex plane. The choice of contour is crucial, and it allows for the evaluation of integrals that might be challenging using real integration methods.

Complex Line Integral

The complex line integral of a function $f(z)$ along a contour C is given by:

$$\oint_C f(z)\,dz$$

This integral is evaluated by parameterizing the contour and performing a real integration.

Example: Basic Contour Integral

Consider the contour integral of the function $f(z) = z^2$ along a simple closed contour, such as a circle.

5.3.2   Cauchy's Integral Formula

Cauchy's Integral Formula establishes a deep connection between the values of a function inside a closed contour and its values on the contour itself. It is a fundamental result in complex analysis.

Cauchy's Integral Formula

For a function $f(z)$ that is analytic within and on a simple closed contour C, the formula is given by:

$$f(a) = \frac{1}{2\pi i} \oint_C \frac{f(z)}{z - a}\,dz$$

This formula relates the value of $f(a)$ to the function values on the contour C.

Example: Applying Cauchy's Formula

Apply Cauchy's Integral Formula to evaluate the integral $\oint_C \frac{e^z}{z-2}\,dz$ along a suitable contour.

5.3.3   Residue Theory

Residue theory is a powerful tool in contour integration for evaluating complex integrals involving singularities.

Residue at a Singularity

The residue of a function $f(z)$ at a singular point z_0 is denoted as $\text{Res}(f, z_0)$ and is calculated using:

$$\text{Res}(f, z_0) = \lim_{z \to z_0} (z - z_0) f(z)$$

Cauchy's Residue Theorem

Cauchy's Residue Theorem states that for a function $f(z)$ that is analytic within and on a simple closed contour C except for isolated singularities, the integral is given by:

$$\oint_C f(z)\, dz = 2\pi i \sum \text{Res}(f, z_k)$$

This theorem simplifies the evaluation of certain complex integrals.

Example: Evaluating Integral with Residue

Use residue theory to evaluate the integral $\oint_C \frac{1}{z^2 - 1}\, dz$ along a suitable contour.

5.3.4   Applications in Signal Processing: Fourier Transform

Contour integration has applications in signal processing, particularly in the evaluation of Fourier transforms.

Example: Contour Integration in Fourier Transform

Illustrate how contour integration can be applied to evaluate a specific Fourier transform integral.

5.3.5   Numerical Examples: Complex Integrals

Explore numerical methods for approximating complex integrals when closed-form solutions are challenging to obtain.

Numerical Integration: Trapezoidal Rule

Apply the trapezoidal rule for numerical integration to approximate a complex integral along a given contour.

5.4   Residue Theorem and Its Applications

The Residue Theorem is a powerful tool in complex analysis that simplifies the evaluation of certain complex integrals. This section explores the Residue Theorem and its applications in engineering mathematics.

5.4.1   Introduction to the Residue Theorem

The Residue Theorem is a consequence of Cauchy's Integral Formula and provides a systematic way to evaluate integrals of functions with singularities.

Residue of a Function

For a function $f(z)$ with an isolated singularity at z_0, the residue is given by:

$$\text{Res}(f, z_0) = \lim_{z \to z_0} (z - z_0) f(z)$$

The Residue Theorem relates the contour integral of $f(z)$ around a closed curve C to the sum of residues inside C:

$$\oint_C f(z)\, dz = 2\pi i \sum \text{Res}(f, z_k)$$

Example: Evaluating Integral Using Residue Theorem

Apply the Residue Theorem to evaluate the integral $\oint_C \frac{1}{z^2 - 1}\, dz$ around a suitable contour.

5.4.2 Applications in Control Systems: Laplace Transform

The Residue Theorem is extensively used in control systems engineering, particularly in the analysis of Laplace transforms.

Laplace Transform and Residue Theorem

The Laplace transform of a function $f(t)$ can be expressed using the Residue Theorem, facilitating the analysis of linear time-invariant systems.

Example: Laplace Transform Analysis

Utilize the Residue Theorem to compute the Laplace transform of a function with singularities and analyze its behavior in the frequency domain.

5.4.3 Pole-Zero Analysis and Filter Design

Pole-zero analysis is a crucial step in filter design. The Residue Theorem helps identify the poles and zeros of a transfer function.

Transfer Function and Residue Theorem

The transfer function of a system can be expressed in terms of poles and zeros using the Residue Theorem.

Example: Filter Design

Design a digital filter by performing pole-zero analysis using the Residue Theorem and analyze its frequency response.

5.4.4 Applications in Electromagnetics: Complex Integration

In electromagnetics, complex integration and the Residue Theorem are employed in the analysis of electric fields and potential.

Electric Field Analysis

Express the electric field using complex potential and apply the Residue Theorem to evaluate electric field integrals.

Example: Electric Field around a Charge Distribution

Determine the electric field around a given charge distribution using complex integration and the Residue Theorem.

5.4.5 Numerical Examples: Engineering Simulations

Explore numerical simulations of complex integrals using methods such as numerical contour integration.

Numerical Contour Integration

Implement numerical techniques to approximate complex integrals with multiple singularities and assess the accuracy of the results.

5.5 Conformal Mappings

Conformal mappings are transformations that preserve angles and locally maintain shapes in the complex plane. This section explores the concept of conformal mappings, their properties, and practical applications in engineering mathematics.

5.5.1 Introduction to Conformal Mappings

A conformal mapping $f(z)$ is a complex function that preserves angles between intersecting curves at a given point. Conformal mappings play a crucial role in various branches of physics and engineering.

Angle Preservation

For a conformal mapping $f(z)$, the angle between two curves at a point z is preserved, meaning that the mapping does not distort the angles.

Example: Conformal Mapping of a Circle

Illustrate a simple conformal mapping that transforms a circle in the complex plane while preserving angles.

5.5.2   Properties of Conformal Mappings

Conformal mappings possess several key properties, including angle preservation, orientation preservation, and the preservation of infinitesimally small shapes.

Orientation Preservation

A conformal mapping preserves the orientation of curves, meaning it does not reverse the direction of traversing a curve.

Example: Preserving Orientation

Demonstrate a conformal mapping that preserves the orientation of a given curve.

Invariance of Infinitesimally Small Shapes

Conformal mappings preserve infinitesimally small shapes, ensuring that the local structure is maintained.

Example: Invariance of Small Shapes

Explore a conformal mapping that preserves the shapes of small elements in the complex plane.

5.5.3 Applications in Heat Conduction: Complex Potential

Conformal mappings find applications in heat conduction problems through the use of complex potentials.

Complex Potential in Heat Conduction

Express the temperature distribution in a conducting material using a complex potential and illustrate how conformal mappings simplify the analysis.

Example: Conformal Mapping in Heat Conduction

Apply a conformal mapping to analyze heat conduction in a non-uniformly shaped material and determine the temperature distribution.

5.5.4 Mapping of Special Regions: Upper Half-Plane

Certain conformal mappings are particularly useful for transforming specific regions in the complex plane.

Mapping to Upper Half-Plane

Explore a conformal mapping that transforms a given region to the upper half-plane, simplifying mathematical analysis.

Example: Transforming a Region

Apply the mapping to transform a complex region and discuss the advantages of working in the upper half-plane.

5.5.5 Numerical Examples: Engineering Simulations

Numerical simulations involving conformal mappings are employed to analyze complex systems.

Numerical Simulation: Fluid Flow

Utilize numerical methods to simulate fluid flow around a complex object, incorporating conformal mappings to simplify the geometry.

Numerical Simulation: Electrical Fields

Simulate the distribution of electrical fields around intricate structures using conformal mappings for effective analysis.

Chapter 6

Probability and Statistics

6.1 Probability Space and Events

Probability theory is a fundamental branch of mathematics used in various engineering applications. This section introduces the concept of a probability space, events, and explores their properties.

6.1.1 Introduction to Probability Space

A probability space is a mathematical construct that formalizes the notion of uncertainty and randomness. It consists of three elements: a sample space S, a set of events $\mathcal{F}$, and a probability measure P.

Sample Space S

The sample space S is the set of all possible outcomes of an experiment. It is denoted as $S = \{s_1, s_2, \ldots, s_n\}$, where s_i represents a distinct outcome.

Set of Events $\mathcal{F}$

The set of events $\mathcal{F}$ is a collection of subsets of the sample space S. Each subset represents a possible event or a combination of outcomes.

Probability Measure P

The probability measure P assigns a probability to each event in $\mathcal{F}$ and satisfies the following properties:

1. $0 \leq P(A) \leq 1$ for any event A. 2. $P(S) = 1$. 3. If $A_1, A_2, \ldots$ are mutually exclusive events, then $P(A_1 \cup A_2 \cup \ldots) = P(A_1) + P(A_2) + \ldots$.

6.1.2   Events and Their Properties

Events are subsets of the sample space and can be classified based on their properties.

Simple Event

A simple event is a single outcome of an experiment, represented by a singleton set.

Compound Event

A compound event is a combination of outcomes, represented by a non-singleton set.

Complementary Event

The complementary event A' of an event A consists of all outcomes in the sample space that are not in A.

Mutually Exclusive Events

Events A and B are mutually exclusive if $A \cap B = \emptyset$, meaning they cannot occur simultaneously.

Independent Events

Events A and B are independent if the occurrence of one does not affect the occurrence of the other, i.e., $P(A \cap B) = P(A) \cdot P(B)$.

6.1.3 Probability Calculations and Examples

Calculating probabilities involves applying the rules of probability to specific events.

Probability of a Simple Event

The probability of a simple event A is given by $P(A)$.

Probability of a Compound Event

The probability of a compound event $A \cup B$ is calculated using $P(A \cup B) = P(A) + P(B) - P(A \cap B)$.

Example: Tossing a Coin

Consider the experiment of tossing a fair coin. Calculate the probability of getting heads (H) and tails (T).

Example: Rolling a Die

In the experiment of rolling a fair six-sided die, find the probability of rolling an even number (E) or a prime number (P).

6.1.4 Conditional Probability

Conditional probability measures the likelihood of an event occurring given that another event has already occurred.

Conditional Probability Formula

The conditional probability of event A given event B is denoted as $P(A|B)$ and is calculated using:

$$P(A|B) = \frac{P(A \cap B)}{P(B)}$$

Example: Conditional Probability

In a deck of cards, find the probability of drawing an Ace (A) given that a spade (S) was drawn.

6.1.5   Bayes' Theorem

Bayes' Theorem relates conditional probabilities and prior probabilities.

Bayes' Theorem Formula

For events A and B with $P(B) > 0$, Bayes' Theorem is given by:

$$P(A|B) = \frac{P(B|A) \cdot P(A)}{P(B)}$$

Example: Bayes' Theorem

Apply Bayes' Theorem to update probabilities based on new information in a medical diagnosis scenario.

6.1.6   Numerical Examples: Engineering Applications

Probability theory finds extensive use in engineering applications, such as reliability analysis and risk assessment.

Reliability Analysis

Utilize probability concepts to assess the reliability of a system composed of multiple components.

Risk Assessment

Apply probability theory to assess and quantify risks associated with engineering projects.

6.2 Random Variables and Probability Distributions

Random variables are essential in modeling uncertain quantities in engineering applications. This section introduces the concept of random variables, probability distributions, and explores their properties.

6.2.1 Introduction to Random Variables

A random variable is a variable whose values depend on the outcome of a random experiment. It can take on various numerical values, each associated with a probability.

Discrete and Continuous Random Variables

Random variables can be classified as discrete or continuous. Discrete random variables take on countable values, while continuous random variables can take on any value within a range.

Probability Mass Function (PMF)

For a discrete random variable X, the probability mass function (PMF) $P(X = x)$ gives the probability that X takes on the value x.

Probability Density Function (PDF)

For a continuous random variable X, the probability density function (PDF) $f(x)$ specifies the likelihood of X falling within a small interval around x.

6.2.2 Common Probability Distributions

Several probability distributions are commonly used to model random variables in engineering.

Discrete Distributions

1. Bernoulli Distribution: Models a binary outcome (success or failure) with parameter p. PMF: $P(X = k) = p^k \cdot (1 - p)^{1-k}$.

2. Binomial Distribution: Represents the number of successes in n independent Bernoulli trials. PMF: $P(X = k) = \binom{n}{k} p^k \cdot (1 - p)^{n-k}$.

3. Poisson Distribution: Models the number of events occurring in a fixed interval of time or space. PMF: $P(X = k) = \frac{\lambda^k e^{-\lambda}}{k!}$.

Continuous Distributions

1. Uniform Distribution: All values in an interval have equal probability. PDF: $f(x) = \frac{1}{b-a}$ for $a \leq x \leq b$.

2. Normal Distribution: Widely used to model real-valued random variables. PDF: $f(x) = \frac{1}{\sqrt{2\pi\sigma^2}} e^{-\frac{(x-\mu)^2}{2\sigma^2}}$.

3. Exponential Distribution: Models the time until an event occurs in a Poisson process. PDF: $f(x) = \lambda e^{-\lambda x}$ for $x \geq 0$.

6.2.3   Expectation and Variance

The expectation (mean) and variance characterize the central tendency and spread of a random variable.

Expectation of a Random Variable

For a discrete random variable X, the expectation μ is given by $\mu = E(X) = \sum x P(X = x)$. For a continuous random variable, it is given by $\mu = E(X) = \int x f(x)\, dx$.

Variance of a Random Variable

The variance σ^2 is a measure of the spread of a random variable and is given by $\sigma^2 = \text{Var}(X) = E((X - \mu)^2)$.

6.2.4   Sampling Distributions

In engineering, we often deal with samples from populations. The sampling distribution characterizes the properties of sample statistics.

Sample Mean Distribution

For a sample mean $\bar{X}$ from a population with mean μ and standard deviation σ, the sampling distribution approaches a normal distribution as the sample size increases.

Central Limit Theorem

The Central Limit Theorem states that the distribution of the sample mean becomes approximately normal as the sample size increases, regardless of the shape of the population distribution.

6.2.5   Applications in Reliability Engineering

Random variables and probability distributions are extensively used in reliability engineering to model and analyze the lifetimes of components and systems.

Reliability Analysis

Utilize probability distributions to model the reliability of components and assess the overall system reliability.

Failure Time Distributions

Model the time until failure of components using appropriate probability distributions and analyze the reliability characteristics.

6.2.6   Numerical Examples: Engineering Applications

Apply the concepts of random variables and probability distributions to solve practical engineering problems.

Example: Bernoulli Trial

Model a series of Bernoulli trials and calculate the probability of a specific number of successes.

Example: Normal Distribution

Use the normal distribution to model the heights of a population and calculate probabilities related to specific height ranges.

6.3   Mean, Median, Mode, and Standard Deviation

Measures of central tendency and dispersion are vital in summarizing and interpreting data. This section covers the mean, median, mode, and standard deviation, along with their applications.

6.3.1   Mean (Average)

The mean, often referred to as the average, is a measure of central tendency and is calculated as the sum of all values divided by the number of values.

Arithmetic Mean

For a set of n values $x_1, x_2, \ldots, x_n$, the arithmetic mean $\bar{x}$ is given by:

$$\bar{x} = \frac{x_1 + x_2 + \ldots + x_n}{n}$$

Weighted Mean

When each value x_i is associated with a weight w_i, the weighted mean is calculated as:

$$\bar{x} = \frac{w_1 x_1 + w_2 x_2 + \ldots + w_n x_n}{w_1 + w_2 + \ldots + w_n}$$

6.3.2   Median

The median is the middle value of a data set when arranged in ascending or descending order. For an odd number of values, it is the middle value; for an even number, it is the average of the two middle values.

6.3.3   Mode

The mode is the value that occurs most frequently in a data set.

6.3.4   Standard Deviation

The standard deviation measures the dispersion or spread of a data set around its mean.

Population Standard Deviation

For a population with values $x_1, x_2, \ldots, x_n$ and mean μ, the population standard deviation σ is given by:

$$\sigma = \sqrt{\frac{(x_1 - \mu)^2 + (x_2 - \mu)^2 + \ldots + (x_n - \mu)^2}{n}}$$

Sample Standard Deviation

For a sample with values $x_1, x_2, \ldots, x_n$ and sample mean $\bar{x}$, the sample standard deviation s is given by:

$$s = \sqrt{\frac{(x_1 - \bar{x})^2 + (x_2 - \bar{x})^2 + \ldots + (x_n - \bar{x})^2}{n - 1}}$$

6.3.5   Applications in Engineering

These statistical measures find applications in various engineering domains.

Quality Control

Use mean and standard deviation in quality control to ensure products meet specified standards.

Data Analysis in Research

Apply mean, median, mode, and standard deviation in analyzing research data to draw meaningful conclusions.

6.3.6   Working Examples

Example: Calculating Mean and Standard Deviation

Consider a dataset of values $10, 15, 20, 25, 30$. Calculate the mean and standard deviation.

Example: Comparing Measures in a Sample

Analyze a sample dataset with values $5, 10, 10, 15, 20, 20, 25$. Calculate the mean, median, mode, and standard deviation.

6.3.7   Numerical Examples: Engineering Applications

Example: Process Control in Manufacturing

In a manufacturing process, measure the mean and standard deviation of product dimensions to control the production process.

Example: Traffic Flow Analysis

Analyze the mean and median traffic flow data to optimize traffic signal timings in a smart city project.

6.4 Probability Density Functions

Probability density functions (PDFs) play a crucial role in continuous probability distributions, providing a framework for understanding the likelihood of different outcomes in continuous random variables.

6.4.1 Introduction to Probability Density Functions

In the realm of continuous random variables, the concept of probability density functions becomes central. Unlike discrete distributions, where we dealt with probabilities assigned to specific values, PDFs deal with ranges of values.

Continuous Random Variables

Continuous random variables, such as time, distance, or temperature, can take on an infinite number of values within a specified range. PDFs help us model the likelihood of observing a particular range of values.

6.4.2 Definition of Probability Density Function

For a continuous random variable X, the probability density function $f(x)$ is a function that satisfies:

$$\int_{-\infty}^{\infty} f(x)\, dx = 1$$

The probability of X lying in the interval $[a, b]$ is given by:

$$P(a \leq X \leq b) = \int_{a}^{b} f(x)\, dx$$

6.4.3 Key Properties of PDFs

Non-Negativity

The PDF is non-negative for all values of x: $f(x) \geq 0$.

Normalization

The total area under the PDF curve is equal to 1: $\int_{-\infty}^{\infty} f(x)\,dx = 1$.

Probability Calculation

The probability of X lying in the interval $[a, b]$ is given by the integral of the PDF over that interval.

6.4.4 Common Probability Density Functions

Uniform Distribution

The PDF of a uniform distribution over the interval $[a, b]$ is given by:

$$f(x) = \frac{1}{b - a} \text{ for } a \leq x \leq b$$

Normal Distribution

The PDF of a normal distribution with mean μ and standard deviation σ is given by:

$$f(x) = \frac{1}{\sqrt{2\pi\sigma^2}} e^{-\frac{(x-\mu)^2}{2\sigma^2}}$$

Exponential Distribution

The PDF of an exponential distribution with rate parameter λ is given by:

$$f(x) = \lambda e^{-\lambda x} \text{ for } x \geq 0$$

6.4.5 Applications in Engineering

Signal Processing

In signal processing, PDFs are used to model the distribution of signal amplitudes, helping engineers design robust communication systems.

Risk Assessment

PDFs are employed in risk assessment to model the uncertainty associated with variables like project completion time or financial returns.

6.4.6 Working Examples

Example: Uniform Distribution

Consider a uniform distribution over the interval $[2, 8]$. Calculate the probability of X lying in the interval $[4, 6]$.

Example: Normal Distribution

Given a normal distribution with $\mu = 10$ and $\sigma = 2$, find the probability of X being in the interval $[8, 12]$.

Example: Exponential Distribution

For an exponential distribution with $\lambda = 0.5$, calculate the probability of X being greater than 3.

6.5 Correlation and Regression Analysis

Correlation and regression analysis are statistical techniques that help us understand the relationship between two or more variables. These methods are essential in engineering for making predictions, optimizing processes, and decision-making.

6.5.1 Correlation Analysis

Pearson's Correlation Coefficient

Pearson's correlation coefficient (r) measures the strength and direction of a linear relationship between two variables, X and Y. It ranges from -1 to 1, where -1 indicates a perfect negative linear relationship, 1 indicates a perfect positive linear relationship, and 0 indicates no linear relationship.

$$r = \frac{\sum (X_i - \bar{X})(Y_i - \bar{Y})}{\sqrt{\sum (X_i - \bar{X})^2 \sum (Y_i - \bar{Y})^2}}$$

Spearman's Rank Correlation Coefficient

Spearman's rank correlation coefficient (ρ) assesses the strength and direction of a monotonic relationship between two variables. It is based on the ranks of the data rather than the actual values.

$$\rho = 1 - \frac{6 \sum d_i^2}{n(n^2 - 1)}$$

6.5.2   Regression Analysis

Simple Linear Regression

Simple linear regression models the relationship between a dependent variable Y and an independent variable X using the equation:

$$Y = \beta_0 + \beta_1 X + \varepsilon$$

where β_0 is the intercept, β_1 is the slope, and ε is the error term.

Multiple Linear Regression

In multiple linear regression, more than one independent variable is considered:

$$Y = \beta_0 + \beta_1 X_1 + \beta_2 X_2 + \ldots + \beta_k X_k + \varepsilon$$

6.5.3   Applications in Engineering

Quality Control

Correlation and regression help analyze the relationship between process variables and product quality, allowing engineers to optimize manufacturing processes.

Performance Prediction

In fields like civil engineering, regression analysis helps predict the performance of materials under different conditions.

6.5.4 Working Examples

Example: Correlation Analysis

Analyze the correlation between the temperature and energy consumption of a building over a month.

Example: Simple Linear Regression

Predict the sales of a product based on advertising spending using simple linear regression.

Example: Multiple Linear Regression

In a chemical process, predict the yield based on temperature, pressure, and reaction time using multiple linear regression.

6.5.5 Numerical Examples: Engineering Applications

Example: Correlation in Structural Engineering

Assess the correlation between the length of a bridge span and the load it can bear in structural engineering.

Example: Regression in Environmental Engineering

Use regression analysis to predict air quality based on pollutant levels, temperature, and humidity.

Chapter 7

Numerical Methods

7.1 Solutions of Algebraic and Transcendental Equations

Numerical methods for solving algebraic and transcendental equations are fundamental in engineering applications where analytical solutions may not be readily available. These methods provide efficient ways to find roots of equations, critical in designing and analyzing systems.

7.1.1 Bisection Method

The bisection method is a simple and robust technique for finding the root of a continuous function within a given interval $[a, b]$. It relies on the intermediate value theorem.

$$f(c) = 0 \text{ where } c \text{ is in } [a, b]$$

The iteration formula is:

$$c_{n+1} = \frac{a_n + b_n}{2}$$

7.1.2   Newton-Raphson Method

The Newton-Raphson method is an iterative technique for finding successively better approximations to the roots of a real-valued function. It requires an initial guess x_0.

$$x_{n+1} = x_n - \frac{f(x_n)}{f'(x_n)}$$

7.1.3   Secant Method

The secant method is an iterative numerical technique for finding the root of a real-valued function. It does not require the computation of the derivative.

$$x_{n+1} = x_n - \frac{f(x_n)(x_n - x_{n-1})}{f(x_n) - f(x_{n-1})}$$

7.1.4   Applications in Engineering

Circuit Design

In electrical engineering, these methods are used to analyze and design circuits by solving complex nonlinear equations representing circuit behavior.

Structural Analysis

Civil and mechanical engineers use numerical methods to find roots of equations related to structural stability and material properties.

7.1.5   Working Examples

Example: Bisection Method

Apply the bisection method to find the root of $f(x) = x^3 - 6x^2 + 11x - 6$ in the interval $[1, 3]$.

Example: Newton-Raphson Method

Use the Newton-Raphson method to find the root of $f(x) = e^x - 4x$ with an initial guess $x_0 = 1$.

Example: Secant Method

Apply the secant method to find the root of $f(x) = \sin(x) - x$ with initial guesses $x_0 = 0$ and $x_1 = \frac{\pi}{4}$.

7.1.6 Numerical Examples: Engineering Applications

Example: Electronic Circuit Design

Use numerical methods to find the operating point of a transistor in an electronic circuit.

Example: Structural Stability

Apply numerical techniques to determine the critical load for a beam in structural engineering.

7.2 Interpolation and Approximation

Interpolation and approximation techniques are essential in engineering for estimating values between known data points or representing complex functions with simpler ones. These methods find applications in fields such as signal processing, curve fitting, and data analysis.

7.2.1 Linear Interpolation

Linear interpolation is a straightforward method for estimating values between two known data points (x_0, y_0) and (x_1, y_1). The interpolated value y at a point x is given by:

$$y = y_0 + \frac{(x - x_0)(y_1 - y_0)}{x_1 - x_0}$$

7.2.2 Lagrange Interpolation

Lagrange interpolation is a polynomial interpolation method that passes through all given data points. For $n + 1$ data points $(x_0, y_0), (x_1, y_1), \ldots, (x_n, y_n)$, the Lagrange polynomial is:

$$P(x) = \sum_{i=0}^{n} y_i \prod_{j=0, j \neq i}^{n} \frac{(x - x_j)}{(x_i - x_j)}$$

7.2.3 Least Squares Approximation

The least squares approximation method minimizes the sum of squared differences between the actual and approximated values. For a set of n data points $(x_0, y_0), (x_1, y_1), \ldots, (x_{n-1}, y_{n-1})$, the linear approximation is given by:

$$y = a_0 + a_1 x$$

The coefficients a_0 and a_1 are determined by minimizing the expression:

$$\sum_{i=0}^{n-1} (y_i - (a_0 + a_1 x_i))^2$$

7.2.4 Applications in Engineering

Signal Processing

Interpolation is used to estimate intermediate values in a signal, helping engineers reconstruct and analyze signals.

Curve Fitting

In civil engineering, least squares approximation is applied to fit curves to measured data, aiding in design and analysis.

7.2.5 Working Examples

Example: Linear Interpolation

Estimate the temperature at 2:30 PM given data points at 2:00 PM (25°C) and 3:00 PM (30°C).

Example: Lagrange Interpolation

Interpolate the value of $f(2.5)$ for $f(x) = x^2 + 2x + 1$ using Lagrange interpolation.

Example: Least Squares Approximation

Approximate a linear function that best fits the data points $(1, 2), (2, 4), (3, 5)$ using the least squares method.

7.2.6 Numerical Examples: Engineering Applications

Example: Digital Image Processing

Use interpolation to upscale an image in digital image processing.

Example: Traffic Flow Analysis

Apply least squares approximation to model and analyze traffic flow data.

7.3 Numerical Integration and Differentiation

Numerical integration and differentiation methods are crucial in engineering for approximating definite integrals and calculating derivatives of functions. These methods provide efficient ways to evaluate mathematical expressions that may be challenging to integrate or differentiate analytically.

7.3.1   Numerical Integration

Trapezoidal Rule

The trapezoidal rule is a simple numerical integration method that approximates the area under a curve. For a function $f(x)$ in the interval $[a, b]$, the integral is approximated as:

$$\int_a^b f(x)\, dx \approx \frac{b - a}{2} \left[f(a) + f(b) \right]$$

Simpson's Rule

Simpson's rule provides a more accurate estimate of the integral by fitting parabolic arcs to the curve. For an even number of intervals n in the interval $[a, b]$, the formula is:

$$\int_a^b f(x)\, dx \approx \frac{b - a}{6n} \left[f(a) + 4f(a + h) + 2f(a + 2h) + \ldots + 4f(b - h) + f(b) \right]$$

7.3.2   Numerical Differentiation

Forward Difference

The forward difference method is a simple numerical technique for approximating the first derivative of a function $f(x)$ at a point x. It is given by:

$$f'(x) \approx \frac{f(x + h) - f(x)}{h}$$

Central Difference

The central difference method provides a more accurate estimate of the first derivative by considering points on both sides of the evaluation point. It is given by:

$$f'(x) \approx \frac{f(x + h) - f(x - h)}{2h}$$

7.3.3   Applications in Engineering

Signal Processing

Numerical integration is used in signal processing to calculate the area under signals, representing energy or power.

Control Systems

In control systems, numerical differentiation is employed to determine rates of change, aiding in system analysis and design.

7.3.4   Working Examples

Example: Trapezoidal Rule

Approximate the integral $\int_0^1 e^{-x^2}\, dx$ using the trapezoidal rule with $n = 4$ intervals.

Example: Simpson's Rule

Estimate the integral $\int_0^{\pi/2} \sin(x)\, dx$ using Simpson's rule with $n = 6$ intervals.

Example: Forward Difference

Approximate the derivative $f'(2)$ for $f(x) = x^2 + 3x + 2$ using the forward difference method with $h = 0.1$.

Example: Central Difference

Calculate the derivative $f'(1)$ for $f(x) = e^x - \sin(x)$ using the central difference method with $h = 0.01$.

7.3.5   Numerical Examples: Engineering Applications

Example: Power Calculation

In electrical engineering, use numerical integration to calculate the power consumed by a circuit.

Example: System Analysis

Apply numerical differentiation to analyze the dynamic response of a control system.

7.4   Solution of Ordinary Differential Equations

Numerical methods play a pivotal role in solving ordinary differential equations (ODEs) in engineering. ODEs describe the rate of change of a function with respect to an independent variable and are encountered in various fields such as physics, chemistry, and engineering.

7.4.1   Euler's Method

Euler's method is a simple numerical technique for solving first-order ODEs. Given an ODE $y' = f(x, y)$ with initial condition $y(x_0) = y_0$, the next value of y at $x_{n+1} = x_n + h$ is approximated as:

$$y_{n+1} = y_n + h \cdot f(x_n, y_n)$$

7.4.2   Runge-Kutta Methods

Runge-Kutta methods provide higher accuracy in solving ODEs compared to Euler's method. The most commonly used is the fourth-order Runge-Kutta method, given by:

$$k_1 = h \cdot f(x_n, y_n)$$

$$k_2 = h \cdot f(x_n + \frac{h}{2}, y_n + \frac{k_1}{2})$$

$$k_3 = h \cdot f(x_n + \frac{h}{2}, y_n + \frac{k_2}{2})$$

$$k_4 = h \cdot f(x_n + h, y_n + k_3)$$

$$y_{n+1} = y_n + \frac{1}{6}(k_1 + 2k_2 + 2k_3 + k_4)$$

7.4.3   Applications in Engineering

Mechanical Engineering

ODEs are used to model and simulate mechanical systems, such as the motion of a pendulum or the behavior of a spring-mass-damper system.

Chemical Engineering

Chemical reaction kinetics can be described using ODEs, and numerical methods help analyze and predict reaction behavior.

7.4.4   Working Examples

Example: Euler's Method

Solve the ODE $y' = x + y$ with the initial condition $y(0) = 1$ using Euler's method over the interval $[0, 1]$ with $h = 0.1$.

Example: Fourth-Order Runge-Kutta

Use the fourth-order Runge-Kutta method to solve the ODE $y' = -2xy$ with the initial condition $y(0) = 1$ over the interval $[0, 2]$ with $h = 0.2$.

Example: Engineering System Dynamics

Model the dynamics of a simple engineering system using ODEs and simulate its behavior over time.

7.4.5 Numerical Examples: Engineering Applications

Example: Structural Analysis

Apply numerical methods to solve ODEs governing the deformation of structures under applied loads.

Example: Heat Transfer

Use ODEs to model heat transfer in engineering applications, and employ numerical methods to predict temperature distributions.

7.5 Finite Difference Methods

Finite difference methods are widely used in numerical mathematics for approximating solutions to differential equations. These methods discretize the differential operators, allowing the solution of problems involving complex geometries or boundary conditions.

7.5.1 Forward Difference

The forward difference formula is commonly used to approximate the first derivative of a function. For a function $f(x)$, the forward difference is given by:

$$f'(x) \approx \frac{f(x+h) - f(x)}{h}$$

7.5.2 Central Difference

The central difference formula provides a more accurate approximation of the first derivative. It considers points on both sides of the evaluation point x:

$$f'(x) \approx \frac{f(x+h) - f(x-h)}{2h}$$

7.5.3   Backward Difference

The backward difference formula is another method for approximating the first derivative, considering points before and after x:

$$f'(x) \approx \frac{f(x) - f(x - h)}{h}$$

7.5.4   Finite Difference Scheme for Heat Equation

Consider the one-dimensional heat equation $u_t = \alpha u_{xx}$ modeling heat transfer in a rod. Discretizing in space and time using central differences, the finite difference scheme is:

$$\frac{u_i^{n+1} - u_i^n}{\Delta t} = \alpha \frac{u_{i+1}^n - 2u_i^n + u_{i-1}^n}{(\Delta x)^2}$$

7.5.5   Applications in Engineering

Structural Analysis

Finite difference methods are employed in structural analysis to simulate the behavior of structures under various loads.

Fluid Dynamics

In fluid dynamics, finite difference techniques are used to model the flow of fluids in pipes, channels, and other geometries.

7.5.6   Working Examples

Example: Forward Difference

Approximate the derivative $f'(2)$ for $f(x) = x^2 + 3x + 2$ using the forward difference method with $h = 0.1$.

Example: Central Difference

Estimate the derivative $f'(1)$ for $f(x) = e^x - \sin(x)$ using the central difference method with $h = 0.01$.

Example: Heat Conduction

Apply the finite difference scheme to simulate heat conduction in a rod over time.

7.5.7   Numerical Examples: Engineering Applications

Example: Structural Deformation

Use finite difference methods to analyze the deformation of a beam under applied loads.

Example: Fluid Flow Simulation

Simulate fluid flow in a pipeline using finite difference techniques for fluid dynamics.

Chapter 8

Transform Theory

8.1 Laplace Transforms

The Laplace transform is a powerful mathematical tool used to simplify the analysis of linear time-invariant systems. It transforms a function of time, often a differential equation, into a function of a complex variable, making it easier to solve.

8.1.1 Definition of the Laplace Transform

For a function $f(t)$, the Laplace transform $\mathcal{L}\{f(t)\}$ is defined as:

$$\mathcal{L}\{f(t)\} = F(s) = \int_0^\infty e^{-st} f(t) dt$$

where s is a complex number.

8.1.2 Properties of Laplace Transforms

Linearity

The Laplace transform is a linear operation, meaning that $\mathcal{L}\{af(t) + bg(t)\} = a\mathcal{L}\{f(t)\} + b\mathcal{L}\{g(t)\}$.

Shifting Theorem

If $\mathcal{L}\{f(t)\} = F(s)$, then $\mathcal{L}\{e^{at}f(t)\} = F(s-a)$.

Derivative Theorem

The Laplace transform of the derivative of a function is given by $\mathcal{L}\{f'(t)\} = sF(s) - f(0)$.

8.1.3   Applications in Engineering

Control Systems

Laplace transforms are extensively used in control systems to analyze and design dynamic systems.

Circuit Analysis

Electrical engineers use Laplace transforms to analyze and solve linear electrical circuits.

8.1.4   Working Examples

Example: Step Function

Find the Laplace transform of the unit step function $u(t) = 1$ for $t \geq 0$ and $u(t) = 0$ for $t < 0$.

Example: Second-Order Differential Equation

Solve the second-order linear differential equation $y''(t) + 3y'(t) + 2y(t) = 0$ using Laplace transforms.

Example: Circuit Analysis

Apply Laplace transforms to analyze the behavior of an electrical circuit with resistors, capacitors, and inductors.

8.1.5   Numerical Examples: Engineering Applications

Example: System Response

Analyze the time response of a control system using Laplace transforms for stability and performance evaluation.

Example: Signal Processing

Apply Laplace transforms in signal processing to analyze and filter signals in communication systems.

8.2   Fourier Series and Fourier Transforms

Fourier series and Fourier transforms are essential tools in engineering for analyzing and representing periodic and non-periodic signals, respectively. They play a crucial role in signal processing, communication systems, and various other applications.

8.2.1   Fourier Series

Definition

For a periodic function $f(t)$ with period T, the Fourier series representation is given by:

$$f(t) = a_0 + \sum_{n=1}^{\infty} \left[a_n \cos\left(\frac{2\pi nt}{T}\right) + b_n \sin\left(\frac{2\pi nt}{T}\right) \right]$$

where a_0 is the average value, and a_n and b_n are the Fourier coefficients.

Fourier Coefficients

The coefficients a_0, a_n, and b_n are calculated as follows:

$$a_0 = \frac{1}{T} \int_0^T f(t) dt$$

$$a_n = \frac{2}{T} \int_0^T f(t) \cos\left(\frac{2\pi n t}{T}\right) dt$$

$$b_n = \frac{2}{T} \int_0^T f(t) \sin\left(\frac{2\pi n t}{T}\right) dt$$

8.2.2 Fourier Transform

Definition

The Fourier transform of a non-periodic function $f(t)$ is given by:

$$F(\omega) = \int_{-\infty}^{\infty} f(t) e^{-j\omega t} dt$$

where ω is the angular frequency.

Inverse Fourier Transform

The inverse Fourier transform recovers the original function $f(t)$ from its Fourier transform:

$$f(t) = \frac{1}{2\pi} \int_{-\infty}^{\infty} F(\omega) e^{j\omega t} d\omega$$

8.2.3 Applications in Engineering

Signal Processing

Fourier series and transforms are fundamental in signal processing for analyzing and filtering signals.

Communication Systems

In communication systems, Fourier transforms are used to modulate and demodulate signals.

8.2.4   Working Examples

Example: Fourier Series

Find the Fourier series representation of the function $f(t) = \begin{cases} 1, & 0 \leq t < \frac{T}{2} \\ -1, & \frac{T}{2} \leq t < T \end{cases}$.

Example: Fourier Transform

Calculate the Fourier transform of the function $f(t) = e^{-|t|}$ using the integral representation.

Example: Signal Reconstruction

Reconstruct a signal from its Fourier coefficients using the inverse Fourier transform.

8.2.5   Numerical Examples: Engineering Applications

Example: Image Processing

Apply Fourier transforms in image processing for image enhancement and compression.

Example: Spectrum Analysis

Use Fourier analysis to analyze the frequency spectrum of a complex signal in spectrum analysis.

8.3   Z-Transforms

The Z-transform is a powerful tool in engineering for analyzing discrete-time signals and systems. It provides a way to transform sequences into functions of a complex variable, facilitating analysis and design in the discrete domain.

8.3.1 Definition of Z-Transform

For a discrete-time sequence $x[n]$, the Z-transform is defined as:

$$X(z) = Z\{x[n]\} = \sum_{n=-\infty}^{\infty} x[n]z^{-n}$$

where z is a complex variable.

8.3.2 Properties of Z-Transforms

Linearity

The Z-transform is a linear operation, meaning that $Z\{ax_1[n] + bx_2[n]\} = aX_1(z) + bX_2(z)$.

Shifting Theorem

If $Z\{x[n]\} = X(z)$, then $Z\{x[n-k]\} = z^{-k}X(z)$.

Multiplication by n

The Z-transform of $nx[n]$ is given by $X'(z) = -z\frac{dX(z)}{dz}$.

8.3.3 Applications in Engineering

Signal Processing

Z-transforms are extensively used in digital signal processing for system analysis and filter design.

Control Systems

In control systems, Z-transforms are employed to analyze discrete-time systems and design digital controllers.

8.3.4 Working Examples

Example: Z-Transform of a Sequence

Compute the Z-transform of the sequence $x[n] = a^n u[n]$, where $u[n]$ is the unit step function.

Example: System Analysis

Use Z-transforms to analyze the behavior of a discrete-time system described by the difference equation.

Example: Digital Filter Design

Design a digital filter using Z-transforms for a specific frequency response.

8.3.5 Numerical Examples: Engineering Applications

Example: Image Compression

Apply Z-transforms in image compression for efficient storage and transmission of digital images.

Example: Digital Control

Implement Z-transforms in digital control systems for precise control of dynamic processes.

8.4 Convolution and Correlation

Convolution and correlation are fundamental operations in signal processing and system analysis. They play a crucial role in understanding the behavior of linear time-invariant systems and in various engineering applications.

8.4.1   Convolution

Convolution is an operation that combines two signals to produce a third signal, representing the response of a system to an input. For discrete-time signals $x[n]$ and $h[n]$, the convolution is defined as:

$$(x * h)[n] = \sum_{k=-\infty}^{\infty} x[k]h[n - k]$$

Convolution is commutative, associative, and distributive.

8.4.2   Correlation

Correlation measures the similarity between two signals. For discrete-time signals $x[n]$ and $y[n]$, the cross-correlation is given by:

$$(x \star y)[m] = \sum_{n=-\infty}^{\infty} x[n]y[n + m]$$

Auto-correlation is a special case where $x = y$.

8.4.3   Properties of Convolution and Correlation

Associativity

Convolution is associative: $x * (h * g) = (x * h) * g$.

Shift Invariance

For a shift m, $(x \star y)[m] = (y \star x)[-m]$.

Correlation and Convolution

Correlation and convolution are related by $x \star y = x * y^*$, where $*$ denotes complex conjugation.

8.4.4 Working Examples

Example: Convolution in Image Processing

Apply convolution to perform image processing tasks, such as blurring or sharpening.

Example: Cross-Correlation in Communications

Use cross-correlation to detect a signal in the presence of noise in communication systems.

Example: System Response

Determine the response of a linear time-invariant system to an input using convolution.

8.4.5 Numerical Examples: Engineering Applications

Example: Audio Signal Processing

Apply convolution to filter an audio signal for noise reduction or special effects.

Example: Pattern Recognition

Use cross-correlation for pattern recognition in computer vision applications.

Chapter 9

Linear Programming

9.1 Formulation of Linear Programming Problems

Linear programming is a mathematical technique used for optimization in various engineering and business applications. In this section, we delve into the formulation of linear programming problems, exploring the process of modeling real-world scenarios into mathematical equations.

9.1.1 Introduction to Linear Programming

Linear programming involves maximizing or minimizing a linear objective function subject to linear equality and inequality constraints. The general form of a linear programming problem is:

$$\text{Maximize } c_1 x_1 + c_2 x_2 + \ldots + c_n x_n$$

$$\text{Subject to } a_{11} x_1 + a_{12} x_2 + \ldots + a_{1n} x_n \le b_1$$

$$a_{21} x_1 + a_{22} x_2 + \ldots + a_{2n} x_n \le b_2$$

$$\vdots$$

$$a_{m1} x_1 + a_{m2} x_2 + \ldots + a_{mn} x_n \le b_m$$

$$x_1, x_2, \ldots, x_n \ge 0$$

where $c_1, c_2, \ldots, c_n$ are coefficients of the objective function, a_{ij} are coefficients of the constraint equations, and $b_1, b_2, \ldots, b_m$ are constants on the right-hand side of the constraints.

9.1.2   Formulation Steps

Define Decision Variables

Identify the decision variables $(x_1, x_2, \ldots, x_n)$ representing quantities to be optimized.

Formulate Objective Function

Create the objective function representing the quantity to be maximized or minimized.

Establish Constraints

Translate real-world constraints into mathematical equations or inequalities.

Non-negativity Constraints

Specify that decision variables must be non-negative ($x_1, x_2, \ldots, x_n \ge 0$).

9.1.3 Working Examples

Example: Production Planning

Formulate a linear programming problem for production planning, optimizing resource allocation.

Example: Investment Portfolio

Model an investment portfolio problem, optimizing returns while adhering to risk constraints.

Example: Transportation Problem

Formulate a linear programming problem for optimizing transportation costs in a supply chain.

9.1.4 Numerical Examples: Engineering Applications

Example: Project Scheduling

Apply linear programming to optimize project scheduling, considering resource constraints.

Example: Network Flow

Model a network flow problem, optimizing the flow of goods or information in a network.

9.2 Basic Concepts of Graphical and Simplex Methods

Graphical and simplex methods are fundamental tools in solving linear programming problems. In this section, we explore the basic concepts of these methods, providing insights into their applications and numerical techniques.

9.2.1   Graphical Method

The graphical method is a visual approach to solving linear programming problems with two decision variables. It involves plotting the constraints on a graph and identifying the feasible region, where the objective function is optimized.

Steps in Graphical Method

1. Identify Decision Variables: Define the decision variables (x_1, x_2) representing quantities to be optimized. 2. Formulate Objective Function: Create the objective function representing the quantity to be maximized or minimized. 3. Plot Constraints: Translate constraints into linear equations and plot them on a graph. 4. Identify Feasible Region: Determine the feasible region, the area where all constraints are satisfied. 5. Optimize Objective Function: Find the point within the feasible region that optimizes the objective function.

Working Example

Consider a production planning problem with two products. Formulate and solve using the graphical method to optimize resource allocation.

9.2.2   Simplex Method

The simplex method is an iterative numerical technique for solving linear programming problems with any number of decision variables. It efficiently identifies the optimal solution by moving from one vertex to another along the edges of the feasible region.

Steps in Simplex Method

1. Formulate Standard Form: Convert the linear programming problem into standard form. 2. Set Up Initial Simplex Tableau: Create the initial simplex tableau based on the standard form. 3. Identify Pivot Element: Determine the pivot column and pivot row to update the tableau. 4. Update Tableau: Perform

row operations to update the tableau and move to the next iteration. 5. Iterate Until Optimality: Continue iterations until an optimal solution is reached.

Working Example

Solve a linear programming problem with three decision variables using the simplex method. Explore the iterations and final optimal solution.

9.2.3 Numerical Examples: Engineering Applications

Example: Manufacturing Process Optimization

Apply the graphical method to optimize a manufacturing process with multiple constraints.

Example: Resource Allocation in Project Management

Utilize the simplex method to optimize resource allocation in a project management scenario.

9.3 Duality and Dual Simplex Method

Duality is a powerful concept in linear programming that establishes a relationship between the primal and dual problems. The dual simplex method is an extension of the simplex method specifically designed for solving the dual problem efficiently.

9.3.1 Duality in Linear Programming

Consider a linear programming problem in standard form:

$$\begin{aligned}
\text{Maximize} \quad & Z = c_1 x_1 + c_2 x_2 + \ldots + c_n x_n \\
\text{Subject to} \quad & a_{11} x_1 + a_{12} x_2 + \ldots + a_{1n} x_n \leq b_1 \\
& a_{21} x_1 + a_{22} x_2 + \ldots + a_{2n} x_n \leq b_2 \\
& \quad \vdots \\
& a_{m1} x_1 + a_{m2} x_2 + \ldots + a_{mn} x_n \leq b_m \\
& x_1, x_2, \ldots, x_n \geq 0
\end{aligned}$$

The corresponding dual problem is given by:

$$\begin{aligned}
\text{Minimize} \quad & W = b_1 y_1 + b_2 y_2 + \ldots + b_m y_m \\
\text{Subject to} \quad & a_{11} y_1 + a_{21} y_2 + \ldots + a_{m1} y_m \geq c_1 \\
& a_{12} y_1 + a_{22} y_2 + \ldots + a_{m2} y_m \geq c_2 \\
& \quad \vdots \\
& a_{1n} y_1 + a_{2n} y_2 + \ldots + a_{mn} y_m \geq c_n \\
& y_1, y_2, \ldots, y_m \geq 0
\end{aligned}$$

The strong duality theorem establishes a relationship between the optimal values of the primal and dual problems: If the primal problem has an optimal solution, so does the dual problem, and the optimal values are equal.

9.3.2 Dual Simplex Method

The dual simplex method is an algorithm designed to solve the dual problem efficiently. It is particularly useful when the primal problem has an infeasible solution or an unbounded optimal solution.

Steps in Dual Simplex Method

1. Start with an initial feasible solution. 2. Use the simplex method to move towards feasibility. 3. Maintain dual feasibility throughout the iterations. 4. Continue iterations until an optimal solution is reached or an infeasibility is detected.

Working Example

Solve a linear programming problem and its dual using the dual simplex method. Explore the iterations and final optimal solution.

9.3.3   Numerical Examples: Engineering Applications

Example: Transportation Network Optimization

Apply the dual simplex method to optimize transportation network costs in an engineering scenario.

Example: Resource Allocation in Project Management

Utilize duality to optimize resource allocation in project management.

Chapter 10

Probability and Statistics (Again)

10.1 Descriptive Statistics

Descriptive statistics is the branch of statistics that deals with the collection, analysis, interpretation, presentation, and organization of data. It provides a summary of the main aspects of a dataset, offering a clear and concise overview.

10.1.1 Measures of Central Tendency

Mean

The mean, denoted by $\bar{x}$, is the sum of all values in a dataset divided by the number of observations:

$$\bar{x} = \frac{\sum_{i=1}^{n} x_i}{n}$$

Median

The median is the middle value in a dataset when it is ordered. If the dataset has an odd number of observations, the median is the middle value. If the dataset has an even number of observations, the median is the average of the two middle values.

Mode

The mode is the value that appears most frequently in a dataset.

10.1.2   Measures of Dispersion

Range

The range is the difference between the maximum and minimum values in a dataset.

Variance

The variance, denoted by σ^2 (population variance) or s^2 (sample variance), measures how far each value in the dataset is from the mean. It is calculated as:

$$\sigma^2 = \frac{\sum_{i=1}^{n}(x_i - \bar{x})^2}{n} \quad \text{(Population)}$$

$$s^2 = \frac{\sum_{i=1}^{n}(x_i - \bar{x})^2}{n - 1} \quad \text{(Sample)}$$

Standard Deviation

The standard deviation is the square root of the variance and provides a measure of the spread of values around the mean.

10.1.3   Working Examples

Consider a dataset of engineering test scores: 85, 92, 88, 78, 95, 90, 87, 92. Calculate the mean, median, mode, range, variance, and standard deviation.

10.1.4 Numerical Examples: Engineering Applications

Example: Quality Control in Manufacturing

Apply descriptive statistics to monitor and control the quality of manufactured products in an engineering context.

Example: Time Analysis in Project Management

Utilize measures of central tendency and dispersion for time analysis in project management.

10.2 Probability Distributions

Probability distributions describe how the probabilities of different values in a dataset are spread. They are essential in engineering for modeling uncertainties and making informed decisions.

10.2.1 Discrete Probability Distributions

Bernoulli Distribution

The Bernoulli distribution models a single experiment with two outcomes: success (1) or failure (0). The probability mass function is given by:

$$P(X = k) = \begin{cases} p & \text{if } k = 1, \\ q = 1 - p & \text{if } k = 0. \end{cases}$$

Binomial Distribution

The binomial distribution models the number of successes in a fixed number of independent Bernoulli trials. The probability mass function is given by:

$$P(X = k) = \binom{n}{k} p^k q^{n-k},$$

where $\binom{n}{k}$ is the binomial coefficient.

Poisson Distribution

The Poisson distribution models the number of events occurring in fixed intervals of time or space. The probability mass function is given by:

$$P(X = k) = \frac{e^{-\lambda}\lambda^k}{k!},$$

where λ is the average rate of occurrence.

10.2.2   Continuous Probability Distributions

Uniform Distribution

The uniform distribution models outcomes with equal probability in a given range. The probability density function is constant within the range.

Normal Distribution

The normal distribution, or Gaussian distribution, is widely used in engineering. The probability density function is given by:

$$f(x) = \frac{1}{\sqrt{2\pi\sigma^2}}e^{-\frac{(x-\mu)^2}{2\sigma^2}},$$

where μ is the mean and σ is the standard deviation.

Exponential Distribution

The exponential distribution models the time between events in a Poisson process. The probability density function is given by:

$$f(x) = \lambda e^{-\lambda x},$$

where λ is the rate of occurrence.

10.2.3   Working Examples

Consider a manufacturing process with a binomial distribution representing defective products. Calculate the probability of a specific number of defects.

10.2.4 Numerical Examples: Engineering Applications

Example: Reliability Engineering

Apply probability distributions to analyze system reliability and failure rates.

Example: Queueing Theory

Utilize probability distributions to model waiting times in queueing systems.

10.3 Statistical Inference

Statistical inference involves drawing conclusions about a population based on a sample. This is crucial in engineering for making predictions, testing hypotheses, and decision-making.

10.3.1 Point Estimation

Mean and Variance

Point estimates aim to provide a single value that is the best guess for an unknown parameter. For the population mean μ and variance σ^2, sample mean $\bar{x}$ and sample variance s^2 are common estimators.

$$\text{Sample Mean: } \bar{x} = \frac{\sum_{i=1}^{n} x_i}{n}$$

$$\text{Sample Variance: } s^2 = \frac{\sum_{i=1}^{n} (x_i - \bar{x})^2}{n - 1}$$

10.3.2 Interval Estimation

Confidence intervals provide a range within which the true parameter is likely to fall with a certain level of confidence.

Confidence Interval for the Mean

For a normal distribution, the confidence interval for the mean μ is given by:

$$\bar{x} \pm z \left(\frac{s}{\sqrt{n}} \right)$$

where z is the critical value.

10.3.3   Hypothesis Testing

Null and Alternative Hypotheses

Hypothesis testing involves making decisions about population parameters based on sample data. The null hypothesis H_0 is typically a statement of no effect, and the alternative hypothesis H_a is what we want to test.

P-value

The p-value is the probability of obtaining results as extreme or more extreme than the observed results under the assumption that the null hypothesis is true.

10.3.4   Working Examples

Consider a scenario where the average strength of a material needs to be estimated from a sample. Use point estimation and confidence intervals for this purpose.

10.3.5   Numerical Examples: Engineering Applications

Example: Quality Control

Apply hypothesis testing in quality control to ensure products meet specified standards.

Example: Reliability Testing

Utilize statistical inference in reliability testing to estimate the lifespan of a product.

10.4 Hypothesis Testing

Hypothesis testing is a critical aspect of statistical inference, providing a structured way to make decisions based on sample data. Engineers often use hypothesis testing to validate assumptions, assess quality, and make informed choices.

10.4.1 Basics of Hypothesis Testing

Null and Alternative Hypotheses

In any hypothesis test, two hypotheses are formulated: the null hypothesis (H_0) and the alternative hypothesis (H_a). The null hypothesis represents the status quo, while the alternative hypothesis suggests a change or effect.

Test Statistic

The test statistic is a numerical summary of the data used to decide whether to reject the null hypothesis. It is often based on sample means, proportions, or variances.

10.4.2 Steps in Hypothesis Testing

1. Formulate the null and alternative hypotheses.

2. Choose the significance level (α) - the probability of rejecting a true null hypothesis.

3. Collect and analyze the sample data.

4. Calculate the test statistic.

5. Make a decision: reject or fail to reject the null hypothesis.

10.4.3   Types of Errors

Type I Error

A Type I error occurs when the null hypothesis is incorrectly rejected. The probability of Type I error is denoted by α.

Type II Error

A Type II error occurs when the null hypothesis is not rejected when it is false. The probability of Type II error is denoted by β.

10.4.4   Working Examples

Consider a scenario where an engineer wants to test if a new manufacturing process improves the strength of a material. Hypothesis testing can be used to make an informed decision.

10.4.5   Numerical Examples: Engineering Applications

Example: Drug Efficacy

Apply hypothesis testing in pharmaceutical engineering to assess the effectiveness of a new drug.

Example: Environmental Impact

Use hypothesis testing to evaluate the environmental impact of a new manufacturing facility.

Chapter 11

Graph Theory

11.1 Basic Concepts

Graph theory is a fundamental mathematical discipline with wide applications in engineering, computer science, and various other fields. In this section, we will explore the basic concepts of graph theory, providing a foundation for more advanced topics.

11.1.1 Graphs and Terminology

A graph, denoted as G, consists of a set of vertices and a set of edges connecting these vertices. The vertices represent entities, and the edges represent relationships between these entities.

Definitions

- **Vertex** (V)**:** Represents an entity.

- **Edge** (E)**:** Represents a relationship between two vertices.

- **Directed Graph:** Edges have a direction.

- **Undirected Graph:** Edges have no direction.

11.1.2   Types of Graphs

Directed Graphs (Digraphs)

In directed graphs, edges have a specific direction, indicating a one-way relationship between vertices.

Undirected Graphs

In undirected graphs, edges have no direction, implying a two-way relationship between vertices.

11.1.3   Working Examples

Example: Social Network Analysis

Consider a social network where individuals are represented as vertices, and friendships are represented as edges. Analyze the network structure.

Example: Transportation Networks

Apply graph theory to model transportation networks, where cities are vertices, and roads are edges.

11.1.4   Numerical Examples: Engineering Applications

Example: Circuit Analysis

Model an electrical circuit using a graph, where components are vertices, and connections are edges.

Example: Project Scheduling

Utilize graph theory to represent project tasks as vertices and dependencies as edges for efficient project scheduling.

11.2 Trees and Their Applications

Trees are a special type of graph with crucial applications in various fields, including network design, data structures, and optimization. In this section, we will delve into the properties of trees and explore their practical applications.

11.2.1 Definition and Properties of Trees

A tree is an undirected graph without cycles. It consists of vertices and edges, meeting the following conditions:

- It is connected, meaning there is a path between every pair of vertices.

- It is acyclic, indicating the absence of cycles or loops.

11.2.2 Types of Trees

Binary Trees

Binary trees are trees in which each vertex has at most two children.

Spanning Trees

A spanning tree of a connected graph is a subgraph that is a tree and spans all the vertices of the original graph.

11.2.3 Working Examples

Example: Network Design

Consider a telecommunication network where vertices represent cities and edges represent communication links. Design an efficient network using trees.

Example: Hierarchical Structures

Model hierarchical structures using trees, such as organizational charts where vertices are employees, and edges represent hierarchical relationships.

11.2.4   Numerical Examples: Optimization

Example: Minimal Spanning Tree

Apply Kruskal's or Prim's algorithm to find the minimal spanning tree in a weighted graph, minimizing the total edge weights.

Example: Huffman Coding

Use trees to implement Huffman coding for data compression, creating efficient binary codes for characters.

11.3   Connectivity and Network Flows

Connectivity and network flows are crucial concepts in graph theory with wide-ranging applications, from transportation networks to information flow. This section explores these concepts and provides practical insights into their applications.

11.3.1   Connectivity in Graphs

Vertex Connectivity

The vertex connectivity of a graph is the minimum number of vertices that need to be removed to disconnect the graph.

Edge Connectivity

Edge connectivity is the minimum number of edges that need to be removed to disconnect the graph.

Menger's Theorems

Menger's theorems provide insights into the connectivity of a graph, relating vertex and edge connectivity.

11.3.2 Network Flows

Network flows deal with the movement of resources through a network. They are essential in optimizing transportation, communication, and resource distribution.

Max-Flow Min-Cut Theorem

The Max-Flow Min-Cut Theorem states that the maximum flow in a network is equal to the minimum capacity of a cut in the network.

Ford-Fulkerson Algorithm

The Ford-Fulkerson algorithm is a method for finding the maximum flow in a network.

11.3.3 Working Examples

Example: Transportation Networks

Optimize the flow of goods in a transportation network by applying network flow algorithms.

Example: Information Flow

Model the flow of information in a communication network, ensuring efficient data transmission.

11.3.4 Numerical Examples

Example: Max-Flow Min-Cut

Apply the Max-Flow Min-Cut Theorem to find the maximum flow in a given network and identify the corresponding minimum cut.

Example: Ford-Fulkerson Algorithm

Use the Ford-Fulkerson algorithm to find the maximum flow in a network with varying edge capacities.

www.ingramcontent.com/pod-product-compliance
Lightning Source LLC
Chambersburg PA
CBHW080947290526
45795CB00009B/2933